# The Marie Curie Collection:

## Writings, Scientific Papers, and Lectures

PIERRE CURIE
AUTOBIOGRAPHICAL NOTES—MARIE CURIE
THE DISCOVERY OF RADIUM
NOBEL PRIZE LECTURE

by Marie Curie

Bamboo & Pine Press

Pierre Curie (Originally published in 1923)
Autobiographical Notes—Marie Curie (Originally published in 1923)
The Discovery Of Radium ( (Originally published in 1921)
Nobel Prize Lecture (Nobel Lecture, December 11, 1911)
by Marie Curie

This edition published 2026 by Bamboo & Pine Press.
Publisher: Bamboo & Pine Press (Zhu & Song Press, LLC)
North Potomac, Maryland 20878
Editor-in-Chief: Xiaohong Zhu
Editorial Contact: editor@zhuandsongpress.com
Cover Design: Zhu & Song Media
Publisher Website: www.zhuandsongpress.com
Printed in: USA
Distribution: Worldwide
ISBN-13: 979-8-88689-029-7
eBook ISBN-13: 979-8-88689-030-3


First Edition

10 9 8 7 6 5 4 3 2 1

# Editor's Introduction

Few scientific figures have left a legacy as profound as that of Marie Curie, a pioneer whose discoveries transformed modern science and medicine. Born in Warsaw in 1867 and later working in Paris, Curie's dedication to scientific research led to groundbreaking work in the study of radioactivity—an area of research that would reshape the fields of physics, chemistry, and medicine. Through perseverance, intellectual rigor, and extraordinary determination, she became one of the most influential scientists in history and the first woman to receive a Nobel Prize.

Marie Curie's work was closely connected with that of her husband and collaborator, Pierre Curie. Together they conducted pioneering experiments on radioactive elements, ultimately discovering two new elements: polonium and radium. Their research not only opened a new frontier in scientific knowledge but also laid the foundations for future developments in nuclear physics and radiochemistry. In recognition of their achievements, Marie and Pierre Curie shared the 1903 Nobel Prize in Physics with Henri Becquerel. Later, in 1911, Marie Curie received the Nobel Prize in Chemistry for her discovery of radium and polonium and for her investigation of the properties of radioactive substances, becoming the first person to win Nobel Prizes in two different scientific fields.

This volume gathers several of Marie Curie's most significant writings and lectures, offering readers insight into both her scientific work and her personal reflections on the discoveries that changed the course of modern science.

The collection opens with Pierre Curie (1923), Marie Curie's deeply personal biography of her husband and scientific partner. In this work she recounts Pierre Curie's life, character, and scientific achievements, while also describing the intellectual collaboration that defined their shared research. The book provides a rare glimpse into the human story behind one of the most important scientific partnerships of the twentieth century.

Following this biographical work are Autobiographical Notes (1923), in which Marie Curie recounts her early life, education, and the path that led her into scientific research. These reflections offer a concise but powerful narrative of her journey from Poland to Paris and illuminate the persistence and dedication required for her scientific achievements.

The collection also includes The Discovery of Radium (1921), a lecture in which Curie recounts the history and scientific significance of the discovery that made her name known throughout the world. In clear and accessible language, she explains the early investigations into radioactivity and the experimental efforts that eventually led to the isolation of radium.

Finally, the volume concludes with Marie Curie's Nobel Lecture, delivered on December 11, 1911. In this lecture she summarizes the scientific discoveries that led to her Nobel Prize in Chemistry and reflects on the broader implications of research into radioactive substances. The lecture stands as both a scientific overview and a testament to the transformative power of careful experimental research.

Together, these writings present not only the scientific achievements of Marie Curie but also the intellectual spirit and perseverance that defined her work. They reveal a scientist deeply committed to the pursuit of knowledge and to the belief that scientific discovery belongs to humanity as a whole.

The texts reproduced in this edition were originally published between 1911 and 1923 and are presented here as part of the public domain record of Marie Curie's writings. In assembling this collection, the aim has been to provide readers with a coherent and accessible introduction to Curie's life, her scientific work, and the historical context in which her discoveries were made.

More than a century after the discovery of radium, Marie Curie's work continues to influence scientific research and medical practice around the world. Her legacy remains a powerful reminder that curiosity, perseverance, and dedication to truth can illuminate new paths of knowledge for generations to come.

# Contents

PIERRE CURIE ........ 1
CHAPTER I ........ 1
CHAPTER II ........ 8
CHAPTER III ........ 14
CHAPTER IV ........ 28
CHAPTER V ........ 40
CHAPTER VI ........ 52
CHAPTER VII ........ 71
AUTOBIOGRAPHICAL NOTES—MARIE CURIE ........ 80
CHAPTER I ........ 80
CHAPTER II ........ 93
CHAPTER III ........ 110
CHAPTER IV ........ 125
THE DISCOVERY OF RADIUM ........ 134
NOBEL PRIZE LECTURE ........ 139

# PIERRE CURIE

(Translated by CHARLOTTE AND VERNON KELLOGG)

## CHAPTER I

### THE CURIE FAMILY. INFANCY AND FIRST STUDIES OF PIERRE CURIE

Pierre Curie's parents, who were educated and intelligent, formed a part of the petite bourgeoisie of small means. They did not frequent fashionable society, but confined themselves entirely to the companionship of their relatives and a few intimate friends.

Eugène Curie, Pierre's father, was a physician and the son of a physician. He knew very few kinsmen of his name, and very little about the Curie family, which was of Alsatian (Eugène Curie was born at Mulhouse in 1827) and Protestant origin. Even though his father was established in London, Eugène had been brought up in Paris, where he pursued his studies in the natural sciences and medicine, and worked as preparator under Gratiollet in the laboratories of the Museum.

Doctor Eugène Curie's remarkable personality impressed all who approached him. He was a tall man, who in youth must have been blonde, with beautiful blue eyes of a clearness and brilliancy that were striking even in an advanced old age. These eyes, which had retained a child-like expression, reflected goodness and intelligence. He had indeed unusual intellectual capacities, a very live aptitude for the natural sciences, and the temperament of a scholar.

Although he wished to consecrate his life to scientific work, family responsibilities following his marriage and the birth of two sons forced him to renounce this desire. The necessities of life obliged him to practice his medical profession. He continued, however, such experimental research as his means permitted, in particular undertaking an investigation upon inoculation for tuberculosis at a time when the bacterial nature of this malady was not yet established. His scientific avocations developed in him the habit of making excursions in search of the plants and animals necessary to his experiments, and this habit, as well as his love of Nature, gave him a marked preference for country life. Until the end of his life he conserved his love for science, and, without doubt, also, his regret at not having been able to devote himself exclusively to it.

His medical career remained always a modest one, but it revealed remarkable qualities of devotion and disinterestedness. At the time of the Revolution of 1848, when he was still a student, the Government of the Republic conferred on him a medal, "for his honorable and courageous conduct" in serving the wounded. He himself had been struck, on February 24th, by a ball which shattered a part of his jaw. A little later, during a cholera epidemic, he installed himself, in order that he might look after the sick, in a quarter of Paris deserted by physicians. During the Commune he established a hospital in his apartment (rue de la Visitation) near which there was a barricade, and there he cared for the wounded. Through this act of civism and because of his advanced convictions he lost a part of his bourgeois patronage. At this time he accepted the position of medical inspector of the organization for the protection of young children. The duties of this post permitted him to live in the suburbs of Paris where health conditions for himself and his family were much better than those of the city.

Doctor Curie had very pronounced political convictions. Temperamentally an idealist, he had embraced with ardor that republican doctrine which inspired the revolutionaries of 1848. He was united in friendship with Henri Brisson and the men of his group. Like them, a free thinker and an anticlerical, he did not have his sons baptized, nor did he have them practice any form of religion.

Pierre's mother, Claire Depouilly, was the daughter of a prominent manufacturer of Puteaux, near Paris. Her father and brothers distinguished themselves through their numerous inventions connected with the making of dyes and special tissues. The family, which was of Savoy, was caught in the business catastrophe caused by the Revolution of 1848, and ruined. And these reverses of fortune, added to those which Doctor Curie had experienced during his career, meant that he and his family lived always in comparatively straightened circumstances, with the difficulties of existence often renewed. Even though raised for a life of ease, Pierre's mother accepted with tranquil courage the precarious conditions which life brought her, and gave proof of an extreme devotion as she made life easier for her husband and children by her activity and her good will.

If the circumstances in which Jacques and Pierre grew up were modest and not free from cares, nevertheless there reigned in the family an atmosphere of gentleness and affection. In speaking to me for the first time of his parents, Pierre Curie said that they were "exquisite." They were, in truth, that. The father's spirit was a little authoritative—always awake and active. And he possessed a rare unselfishness. He neither wished nor knew how to profit by personal relations to ameliorate his condition. He loved his wife and sons tenderly, and was ever ready to aid all who needed him. The mother was slight, vivid in character, and, even though her health had suffered through the birth of her sons, was always gay and active in the simple home that she so well knew how to make attractive and hospitable.

When I first knew them they lived at Sceaux, rue des Sablons (to-day rue Pierre Curie) in a little house of ancient construction half concealed amidst the verdure of a pretty garden. Their life was peaceful. Doctor Curie went where his duties called him, either in Sceaux or in neighboring localities. Beyond this he occupied himself with his garden or his reading. Near relatives and neighbors came to visit on Sundays, when bowling and chess were the favorite amusements. From time to time Henri Brisson sought out his old companion in his tranquil retreat. Great calm and serenity enveloped the garden, the dwelling, and its inhabitants.

Pierre Curie was born the 15th of May, 1859, in a house facing the Jardin des Plantes, rue Cuvier, where his parents lived at the time when his father was working in the Museum laboratories. He was the second son of Doctor Curie and three and a half years younger than his brother Jacques. In after life he retained few particularly characteristic memories of his childhood in Paris; yet he did tell me how vividly present in his mind were the days of the Commune, the battle on the barricade so near the house where he then lived, the hospital established by his father, and the expeditions, on which his brother accompanied him, in search of the wounded.

It was in 1883 that Pierre moved with his parents from the capital to the suburbs of Paris, living first, from 1883 to 1892, at Fontenay-aux-Roses, then at Sceaux from 1892 to 1895, the year of our marriage.

Pierre passed his childhood entirely within the family circle; he never went to the elementary school nor to the lycée. His earliest instruction was given him first by his mother and was then continued by his father and his elder brother, who himself had never followed in any complete way the course of the lycée. Pierre's intellectual capacities were not those which would permit the rapid assimilation of a prescribed course of studies. His dreamer's spirit would not submit itself to the ordering of the intellectual effort imposed by the school. The difficulty he experienced in following such a program was usually attributed to a certain slowness of mind. He himself believed that he had this slow mind and often said so. I think, however, that this belief was not entirely justified. It seems to me, rather, that already from his early youth it was necessary for him to concentrate his thought with great intensity upon a certain definite object, in order to obtain a precise result, and that it was impossible for him to interrupt or to modify the course of his reflections to suit exterior circumstances. It is clear that a mind of this kind can hold within itself great future possibilities. But it is no less clear that no system of education has been especially provided by the public school for persons of this intellectual category, which nevertheless includes more representatives than one would believe at first sight.

Very fortunately for Pierre, who could not, as we can see, become a brilliant pupil in a lycée, his parents had a sufficiently keen intelligence to understand his difficulty, and they refrained from demanding of their son an effort which would have been prejudicial to his development. If, then, Pierre's earliest instruction was irregular and incomplete, it had the advantage of not so weighing on his intelligence as to deform it by dogmas, prejudices or preconceived ideas. And he was always grateful to his parents for this very liberal attitude. He grew up in all freedom, developing his taste for natural science through his excursions into the country, where he collected plants and animals for his father. These excursions, which he made either alone or with one of the family, helped to awake in him a great love of Nature, a passion which endured to the end of his life.

Intimate contact with Nature, which, because of the artificial conditions of city life and of traditional education, few children can know, had a decisive influence on Pierre's development. Guided by his father, he learned to observe facts and to interpret them correctly. He became familiar with the animals and plants of the environs of Paris. He knew which ones could be found at each season of the year in the forests and fields, the streams and ponds. The ponds in particular had for him an ever new attraction with their characteristic vegetation and their population of frogs, tritons, salamanders, dragonflies, and other denizens of air and water. No efforts to obtain the objects of his interests seemed too great for him. He never hesitated to take any animal in his hands in order to examine it more closely. Later, after our marriage, in our walks together, if I made some objection to letting him put a frog into my hands, he would exclaim: "But no, see how pretty it is!" He loved always, too, to bring back bouquets of wild flowers from his walks.

Thus his knowledge of natural history progressed rapidly. At the same time, also, he was mastering the elements of mathematics. His classical studies, on the contrary, had been much neglected, and it was principally through general reading that he acquired a knowledge of literature and history. His father, who was widely cultured, possessed a library containing many French and foreign works. Having himself a very pronounced taste for reading, he was able to communicate it to his son.

When he was about fourteen years old, a very happy event occurred in Pierre's education. He was put under an excellent professor, A. Bazille, who taught him elementary and advanced mathematics. This master was able to appreciate his young pupil, became much attached to him, and directed his work with the greatest solicitude. He even helped him to advance in his study of Latin, in which he was very much behind. At the same time Pierre and Albert Bazille, his professor's son, became friends.

This teaching had, I am sure, a great influence on the mind of Pierre, aiding him to develop and to sound the depth of his faculties and to realize his capacities for science. He had a remarkable aptitude for mathematics, which expressed itself chiefly by a characteristic geometric spirit and a great power of spatial vision. He, therefore, progressed rapidly and joyfully in his studies under M. Bazille, for whom he always felt an unalterable gratitude.

He once told me something which proved that even at this time he was not content solely to follow a fixed program of studies, but that he had already begun to launch out into personal investigation. Strongly attracted by the theory of determinants, which he had just mastered, he undertook to realize an analogous conception, but in three dimensions, and endeavored to discover the properties and uses of these "cubical determinants." Needless to say that at his age, and with the knowledge then at his disposal, such an enterprise was beyond his powers. The attempt, however, was none the less indicative of his awakening inventive spirit.

Several years later, when preoccupied with reflections upon symmetry, he asked himself the question: "Could not one find a general method for the solution of any equation whatever? Everything is a question of symmetry." He did not then know of Galois' theory of groups which had made it possible to attack this problem. But he was happy later to learn its results in the geometric applications to the case of equations of the 5th degree.

Thanks to his rapid progress in mathematics and physics, Pierre Curie was made a bachelor of science at the age of sixteen years. With this he passed his most difficult stage of formal education. The only thing with which he had to concern himself in the future was the acquisition of knowledge through his personal and independent effort in a field of science freely chosen.

# CHAPTER II

## DREAMS OF YOUTH. FIRST SCIENTIFIC WORK. DISCOVERY OF PIEZO-ELECTRICITY

Pierre Curie was still very young when he began his higher studies in preparation for the licentiate in physics. He followed the lectures and laboratory work at the Sorbonne and had, besides, access to the laboratory of Professor Leroux in the School of Pharmacy, where he assisted in the preparation of the physics courses. At the same time he became further acquainted with laboratory methods by working with his brother Jacques, who was then preparator of chemistry courses under Riche and Jungfleisch.

Pierre received his licentiate in physical sciences at the age of eighteen. During his studies he had attracted the attention of Desains, director of the University laboratory, and of Mouton, assistant director of the same laboratory. Thanks to their appreciation he was appointed, when only nineteen years old, preparator for Desains and placed in charge of the students' laboratory work in physics. He held this position five years, and it was during this time that he began his experimental research.

It is to be regretted that because of his financial situation Pierre was obliged, at this early age of nineteen, to accept the post of preparator instead of being able to give his whole time for two or three years longer to his University studies. With his time thus absorbed by his professional duties and his investigations he had to give up following the lectures in higher mathematics, and he therefore passed no further examinations. In compensation, however, he was released from military service in conformity with the privileges at that time accorded young men who undertook to serve as teachers in the public-school system.

He was by this time a tall and slender young man with chestnut-colored hair and a shy and reserved expression. At the same time his youthful face mirrored a profound inner life. One has such an impression of him as he appears in a good group photograph of Doctor Curie's family. His head is resting on his hand in a pose of abstraction and reverie, and one cannot but be struck by the expression of the large, limpid eyes that seem to be following some inner vision. Beside him the brown-haired brother offers a striking contrast, his vivacious eyes and whole appearance suggesting decision.

The two brothers loved each other tenderly and lived as good comrades, being accustomed to work together in the laboratory and walk together in their free hours. They also kept up affectionate relations with a few of their childhood friends: Louis Depouilly, their cousin, who became a physician; Louis Vauthier, also later a physician; and Albert Bazille, who became an engineer in the post and telegraph service.

Pierre used to tell me of the vivid memories he had of the vacations passed at Draveil on the Seine, where, with his brother Jacques, he took long walks beside the river, agreeably interrupted by swimming and diving in the stream. Both brothers were excellent swimmers. Sometimes they tramped for entire days. They had, at an early age, acquired the habit of visiting the suburbs of Paris on foot. At times also Pierre made solitary excursions which well suited his meditative spirit. On these occasions he lost all sense of time, and went to the extreme limit of his physical forces. Absorbed in delightful contemplation of the things about him, he was not conscious of material difficulties.

On the pages of a diary written in 1879,[1] he thus expressed the salutary influence of the country upon him:

"Oh, what a good time I have passed there in that gracious solitude, so far from the thousand little worrying things that torment me in Paris. No, I do not regret my nights passed in the woods, and my solitary days. If I had the time I would let myself recount all my musings. I would also describe my delicious valley, filled with the perfume of aromatic plants, the beautiful mass of foliage, so fresh and so humid, that hung over the Bièvre, the fairy palace with its colonnades of hops, the stony hills, red with heather, where it was so good to be. Oh, I shall remember always with gratitude the forest of the Minière; of all the woods I have seen, it is this one that I have loved most and where I have been happiest. Often in the evening I would start out and ascend again this valley, and I would return with twenty ideas in my head."

Thus, for Pierre Curie, the sensation of well-being he experienced in the country was derived from the opportunity for tranquil reflection. Daily life in Paris with its numerous interruptions did not permit of undisturbed concentration, and this was to him a cause of inquietude and suffering. He felt himself destined for scientific research; for him the necessity was imperative of comprehending the phenomena of Nature in order to form a satisfactory theory to explain them. But when trying to fix his mind on some problem he had frequently to turn aside because of the multiplicity of futile things that disturbed his reflections and plunged him into discouragement.

Under the heading, "A day like too many others," he enumerated in his diary a list of the puerile happenings that had completely filled one of his days, leaving no time for useful work. He then concluded: "There is my day, and I have accomplished nothing. Why?" Further on he returns to the same theme under a title borrowed from Victor Hugo's "Le Roi S'Amuse,"

"To deafen with little bells the spirit that would think."

"In order that, weak one that I am, I shall not let my head turn with all the winds, yielding to the least breath that touches it, it is necessary that all should be immobile about me, or that, like a spinning top, movement alone should render me insensible to external objects.

"When, in the process of turning slowly upon myself, I try to gain momentum, a nothing, a word, a story, a paper, a visit stops me and is able to put off or retard forever the moment when, granted a sufficient swiftness I might have, in spite of my surroundings, concentrated on my own intention.... We must eat, drink, sleep, be idle, love, touch the sweetest things of life and yet not succumb to them. It is necessary that, in doing all this, the higher thoughts to which one is dedicated remain dominant and continue their unmoved course in our poor heads. It is necessary to make a dream of life, and to make of a dream a reality."

This acute analysis, sufficiently surprising in a young man of twenty years, suggests in an admirable manner the conditions necessary to the highest manifestations of the intellect. It carries a lesson which, if it were sufficiently understood, would facilitate the way of all contemplative spirits capable of opening new paths for humanity.

The unity of thought toward which Pierre Curie strove was troubled not only by professional and social obligations but also by his tastes, which urged him towards a broad literary and artistic culture. Like his father, he loved reading, and did not fear to undertake arduous literary tasks. To some criticism made in this connection, he responded readily: "I do not dislike tedious books." This meant that he was fascinated by the search after truth which is sometimes associated with writing devoid of charm. He also loved painting and music, and went gladly to look at pictures or to attend a concert. A few fragments of poetry in his handwriting were left among his papers.

But all these preoccupations were subordinated in his mind to what he considered his true task, and when his scientific imagination was not in full activity, he felt himself, in a sense, an incomplete being. He expressed this inquietude with an emotion born of his suffering during momentary periods of depression.

"What shall I become?" he wrote. "Very rarely have I command of all myself; ordinarily a part of me sleeps. My poor spirit, are you then so weak that you cannot control my body? Oh, my thoughts, you count indeed for very little! I should have the greatest confidence in the power of my imagination to pull me out of the rut, but I greatly fear that my imagination is dead."

But despite hesitations, doubts, and lost moments, the young man was little by little striking out his path and strengthening his will. He was resolutely carrying on fruitful investigations at an age when many men who were to become savants were as yet only pupils.

His first work, done in collaboration with Desains, concerned the determination of the lengths of heat waves with the aid of a thermoelectric element and a metallic wire grating, a process, then entirely new, which has since often been employed in the study of this question.

Following this he undertook an investigation on crystals in collaboration with his brother, who had passed his licentiate and was preparator for Friedel in the laboratory of mineralogy at the Sorbonne. Their experiments led the two young physicists to a great success: the discovery of the hitherto unknown phenomena of piezo-electricity, which consists of an electric polarization produced by the compression or the expansion of crystals in the direction of the axis of symmetry. This was by no means a chance discovery. It was the result of much reflection on the symmetry of crystalline matter, which enabled the brothers to foresee the possibilities of such polarization. The first part of the investigation was made in Friedel's laboratory. With an experimental skill rare at their age, the young men succeeded in making a complete study of the new phenomenon, established the conditions of symmetry necessary to its production in crystals, and stated its remarkably simple quantitative laws, as well as its absolute magnitude for certain crystals. Several well-known scientists of other nations (Roentgen, Kundt, Voigt, Riecke) have made further investigations along this new road opened by Jacques and Pierre Curie.

The second part of the work, and much more difficult to realize experimentally, concerned the compression resulting in piezo-electric crystals when they are exposed to the action of an electric field. This phenomenon, foreseen by Lippmann, was demonstrated by the Curie brothers. The difficulty of the experiment lay in the minuteness of the deformations that had to be observed. Fortunately Desains and Mouton placed a small room adjoining the physics laboratory at the disposal of the brothers so that they might proceed successfully with their delicate operations.

From these researches, as much theoretical as experimental, they immediately deduced a practical application, in the form of a new apparatus, a piezo-electric quartz electrometer, which measures in absolute terms small quantities of electricity, as well as electric currents of low intensity. This apparatus has since then rendered great service in experiments in radioactivity.[2]

During the course of their experiments on piezo-electricity the Curies were obliged to employ electrometric apparatus, and, not being able to use the quadrant electrometer known at that time, they developed a new form of that instrument, better adapted to their necessities. This became known in France as the Curie electrometer. Thus these years of collaboration between the two brothers, always intimately united, proved both happy and fruitful. Their devotion and their common interest in science were to them both a stimulant and a support. During their work the vivacity and energy of Jacques were of precious aid to Pierre, always more easily absorbed by his thoughts.

However, this beautiful and close collaboration lasted only a few years. In 1883, Pierre and Jacques were obliged to separate; Jacques left for the University of Montpellier as Head Lecturer in Mineralogy (Maître de Conferences). Pierre was made Director of Laboratory Work in the School of Industrial Physics and Chemistry founded by the city of Paris at the suggestion of Friedel and of Schützenberger, who became its first director. Their remarkable researches with crystals won for the brothers in 1895—very late, it is true—the Planté prize.

[1]Pierre Curie did not leave a veritable diary but only a few pages as chance permitted, covering but a short period of his life.

[2]The piezo-electric property of quartz has recently had an important application; it has been utilized by P. Langevin in the production of elastic waves of high frequency (beyond sound) sent out in water with the aim of detecting submarine obstacles. This same method can serve in a more general manner to explore ocean depths. We see, here, once again, how pure speculation can lead to discoveries that will be useful later in unforeseen directions.

# CHAPTER III

## LIFE AS THE DIRECTOR OF LABORATORY WORK IN THE SCHOOL OF PHYSICS AND CHEMISTRY. GENERALIZATION OF THE PRINCIPLE OF SYMMETRY. INVESTIGATIONS OF MAGNETISM

It was in the School of Physics, in the old buildings of the Collège Rollin, that Pierre Curie was destined to work, first as Director of Laboratory Work, then as Professor, for twenty-two years, a period covering practically the whole of his scientific life. His memory seemed to cling to these old buildings, now destroyed, in which he had passed all his days, returning only in the evening to his parents in the country. He counted himself fortunate since he enjoyed the favor of the Founder-Director Schützenberger, and the esteem and good will of his students, many of whom became his disciples and friends. In alluding to this experience, at the close of an address delivered at the Sorbonne near the end of his life, he said:

"I desire to recall here that we have made all our investigations in the School of Physics and Chemistry of the city of Paris. In all creative scientific work the influence of the surroundings in which one works is of great importance, and a part of the result is due to that influence. For more than twenty years I have worked in the School of Physics and Chemistry. Schützenberger, the first director of the School, was an eminent scientist. I remember with gratitude that he procured for me opportunities for my own investigations when I was yet but an assistant. Later, he permitted Madame Curie to work beside me, an authorization which was at that time far from an ordinary innovation.

"Schützenberger allowed us all great liberty; his direction made itself felt chiefly through his inspiring love of science. The professors of the School of Physics and Chemistry, and the students who have gone from it, have created a kindly and stimulating atmosphere that has been extremely helpful to me. It is among the old students of the school that we have found our collaborators and our friends. I am happy to be able, here, to thank them all."

The newly appointed director of the laboratory was, when he first assumed his duties, scarcely older than his students, who loved him because of his extreme simplicity of manner, which was much more that of a comrade than of a master. Some of these students recall with emotion their work carried on with him and his discussion at the blackboard, where he readily allowed himself to be led to debate scientific matters to their great profit both in information and in kindled enthusiasm. At a dinner given in 1903 by the Association of Former Students of the School, which he attended, he laughingly recalled an incident of this period. One day after lingering late with several students in the laboratory, he found the door locked, and they all had to climb down from the first floor single file, along a pipe that ran near one of the windows.

Because of his reserve and shyness he did not make acquaintances easily, but those whose work brought them near him loved him because of his kindliness. This was true of his subordinates during his entire life. In the school his laboratory helper, whom he had aided under trying circumstances, thought of him with the greatest gratitude, in fact, with veritable adoration.

Although separated from his brother, he remained bound to him by their former bond of love and confidence. During vacations, Jacques Curie would come to him that they might renew again that valuable collaboration to which both willingly sacrificed their periods of liberty. At times it was Pierre who joined Jacques, who was engaged in making a geological chart of the Auvergne country, and there they covered together the daily distances necessary to the tracing of such a map.

Here are a few memories of these long walks, extracts from a letter written to me shortly before our marriage:

"I have been very happy to pass a little time with my brother. We have been far from all immediate care, and so isolated by our manner of living that we have not even been able to receive a letter, never knowing one night where we would sleep the next. At times it seemed to me that we had gone back to the days when we lived entirely together. Then we always arrived at the same opinions about all things, with the result that it was no longer necessary for us to speak in order to understand each other. This was all the more astonishing because we differed so entirely in character."

From the point of view of scientific investigation, one must recognize that the nomination of Pierre Curie to the School of Physics and Chemistry retarded from the very first his experimental research. Indeed, at the time of his appointment nothing yet existed in that establishment; everything had to be created. Even the walls and the partitions were hardly yet in place. He had, therefore, to organize completely the laboratory and its work, and he acquitted himself of this task in a remarkable manner, injecting into it the spirit of precision and originality so characteristic of him.

The direction of the laboratory work of the large number of students (thirty by promotion) was alone a strain on a young man, assisted as he was only by one laboratory helper. The first years were, therefore, hard years of assiduous work, of benefit chiefly to the students trained and developed by the young laboratory director.

He himself profited by this enforced interruption of his experimental research by trying to complete his scientific studies and, in particular, his knowledge of mathematics. At the same time he became engrossed in considerations of a theoretical nature on the relations between crystallography and physics.

In 1884 he published a memoir on questions of the order and repetition that are at the base of the study of the symmetry of crystals. This was followed in the same year by a more general treatment of the same subject. Another article on symmetry and its repetitions appeared in 1885. In that year he published, too, a very important theoretical work[3] on the formation of crystals, and the capillary constants of the different faces.

This rapid succession of investigations shows how completely engrossed Pierre Curie was in the subject of the physics of crystals.

Both his theoretical and his experimental research in this domain grouped itself around a very general principle, the principle of symmetry, that he had arrived at step by step, and which he only definitely enunciated in memoirs published between the years 1893 and 1895.

The following is the form, already classic, in which he made his announcement:

"When certain causes produce certain effects, the elements of symmetry in the causes ought to reappear in the effects produced.

"When certain effects reveal a certain dissymmetry, this dissymmetry should be apparent in the causes which have given them birth.

"The converse of these two statements does not hold, at least practically; that is to say, the effects produced can be more symmetrical than their causes."

The capital importance of this statement, perfect in its simplicity, lies in the fact that the elements of symmetry which it introduces are related to all the phenomena of physics without exception.

Guided by an exhaustive study of the groups of symmetry which might exist in nature, Pierre Curie showed how one should use this revelation in character at once geometric and physical, in order to foresee whether a particular phenomenon can reproduce itself, or whether its reproduction is impossible under the given conditions. At the beginning of a certain memoir, he insists in these terms:

"I think it is necessary to introduce into physics the ideas of symmetry familiar to crystallographers."

His work in this field is fundamental, and even though he was led away from it later by other investigations, he always retained a passionate interest in the physics of crystals, as well as in projects of further research in this domain.

The principle of symmetry to which Pierre Curie had so eagerly devoted himself is one of the small number of great principles which dominate the study of the phenomena of physics, and which, having their root in ideas derived by experiment, yet little by little detach themselves and assume a form more and more general and more and more perfect. It is in this way that the idea of the equivalence of heat and of work, added to the earlier notion of the equivalence of kinetic and potential energies, brought about the establishment of the principle of the conservation of energy whose application is entirely general. In the same way the law of the conservation of mass grew out of the experiments of Lavoisier, which belong to the foundations of chemistry. Recently an admirable synthesis has made it possible for us to attain a still higher degree of generalization through the union of these two principles, for it has been proved that the mass of a body is proportional to its internal energy. The study of electrical phenomena led Lippmann to announce the general law of the conservation of electricity. The principle of Carnot, born of considerations on the functioning of thermal machines, has acquired also so general a significance, that it made possible the foreseeing of the most probable character of spontaneous evolution for all material systems.

The principle of symmetry furnishes an example of an analogous evolution. To begin with, observation of Nature was able to suggest the idea of symmetry; though such observations reveal only imperfectly any regular dispositions in the aspects of animals and plants. The regularity becomes very much more perfect in the case of crystallized minerals. We may consider that Nature furnishes us the idea of a plane of symmetry and of an axis of symmetry. An object possesses a plane of symmetry, or a plane of reflection, if this plane divides the object into two parts, of which each one may be thought of as the image of the other reflected in the plane as in a mirror. It is this, approximately, that occurs in the external appearance of man and of numerous animals. An object possesses an axis of symmetry of the order n, if it preserves the same appearance after a rotation on this axis of the nth part of a revolution. Thus a regular flower of four petals has an axis of symmetry of the order four, or a quarternary axis. Crystals like those of rock salt or of alum possess many planes of symmetry and many axes of symmetry of different orders.

Geometry teaches us to study the elements of symmetry of a limited figure such, for instance, as a polyhedron; and to discover the relations between its parts which permit us to reunite different symmetries in groups. The knowledge of these groups is of the greatest usefulness in establishing a rational classification of crystal forms in a small number of systems each of which is derived from a simple geometric form. Thus the regular octahedron belongs to the same system as the cube, for in the case of each the group formed by the axes and the planes of symmetry is the same.

In the study of the physical properties of crystalline matter it is necessary to take account of the symmetry of such matter. This is, in general, anisotropic; that is to say, it has not the same properties in all directions. On the other hand, media such as glass or water are isotropic, having equivalent properties in all directions. It was the study of optics which first showed that the propagation of light in a crystal is dependent upon the elements of symmetry in that crystal. The same thing is true for the conduction of heat or electricity, for magnetization, for polarization, etc.

It was in reflecting upon the relations between cause and effect that govern these phenomena that Pierre Curie was led to complete and extend the idea of symmetry, by considering it as a condition of space characteristic of the medium in which a given phenomenon occurs. To define this condition it is necessary to consider not only the constitution of the medium but also its condition of movement and the physical agents to which it is subordinated. Thus a right circular cylinder possesses a plane of symmetry perpendicular to its axis in its position, and an infinity of planes of symmetry pass through its axis. If the same cylinder is in rotation on its axis, the first plane of symmetry persists, but all the others are suppressed. Furthermore, if an electric current traverses the cylinder lengthwise, no plane of symmetry remains.

In every phenomenon the elements of symmetry compatible with its existence may be determined. Certain elements can coexist with certain phenomena, but they are not necessary to them. That which is necessary is that certain ones among these elements shall not exist. It is dissymmetry that creates the phenomenon. When several phenomena are superposed in the same system, the dissymmetries are added together. "Works of Pierre Curie," page 127.

It was from the above considerations that Pierre Curie announced the general law whose text, already cited, attains the highest degree of generalization. The synthesis thus obtained seems complete, and all that was further needed was to deduce from it all the developments of which it admits.

For this it is convenient to define the particular symmetry of each phenomenon and to introduce a classification which makes clear the principal groups of symmetry. Mass, electric charge, temperature, have the same symmetry, of a type called scalar, that of the sphere. A current of water and a rectilineal electric current have the symmetry of an arrow, of the type polar vector. The symmetry of an upright circular cylinder is of the type tensor. All of the physics of crystals can be expressed in a form in which the particular phenomena in question are not specified, but in which are examined only the geometrical and analytical relations between the types of quantities where certain ones are considered as causes and the other as effects.

Thus, the study of electrical polarization by the application of an electric field becomes the examination of the relation between two systems of vectors, and the writing out of a system of linear equations having 9 coefficients. The same system of equations holds for the relation between an electric field and an electric current in crystalline conductors; or for that between the temperature gradient and the heat current, except that the meaning of the coefficients must be changed. Similarly, a study of the general relations between a vector and a system of tensors can reveal all the characteristics of piezo-electric phenomena. And all the rich variety of the phenomena of elasticity depends on the relation between two sets of tensors which require, in principle, 36 coefficients.

The foregoing brief exposition reveals the high philosophic import of these conceptions of symmetry which touch all natural phenomena, and whose profound significance Pierre Curie so clearly set forth. It is interesting in this connection to recall the relation which Pasteur saw between these same conceptions and the manifestations of life. "The universe," he said, "is a dissymmetric whole. I am led to believe that life, as it is revealed to us, must be a function of the dissymmetry of the universe, or of the consequences that it involves."

As his organization of his work in the School progressed, Pierre Curie could begin to dream of going forward again with his experimental research. He could do so, however, only under most precarious conditions, for he had not even a laboratory for his personal work, nor a room of any kind entirely at his disposition. Besides, he possessed no funds to support his investigations. It was only after he had been several years at the School that he obtained, thanks to the influence of Schützenberger, a small annual subvention for his work. Up to that time the materials necessary for him were provided, thanks to the kindness of his superiors, to the extent possible, by drawing upon a very limited general fund of the teaching laboratory. As for a place to work in, he had to content himself with very little. He set up certain of his experiments in the rooms of his pupils when these were not in use. But more frequently he worked in an outside corridor running between a stairway and a laboratory. It was there that he conducted, in particular, his long research on magnetism.

This abnormal state of affairs was manifestly prejudicial to his work, but it had, nevertheless, the happy result of bringing his students closer to him, for it allowed them, at times, to share in his personal scientific interests.

His return to experimental research is marked by a profound study of the "direct reading periodic precision balance for least weights." (1889, 1890, 1891.) In this balance, the use of small weights is suppressed by the employment of a microscope by means of which one reads a micrometer attached to the extremity of one of the arms of the balance. The reading is made when the oscillation of the balance is arrested, which can occur very rapidly, thanks to the use of pneumatic dampeners conveniently constructed. This balance marks a considerable advance over old systems. It has shown itself particularly valuable in laboratories for chemical analysis, where the rapidity of the weighings is frequently a test of precision. We can say that the introduction of the Curie balances marks an epoch in the construction of these instruments. The work done in this field was far from empirical; it comprised a study of the theory of damped movements and the construction of numerous curves established with the aid of some of his students.

It was toward 1891 that Pierre Curie began a long series of investigations on the magnetic properties of bodies at divers temperatures, from the normal up to 1400° C. These investigations, covering years, were presented as a Doctor's thesis before the Faculty of Sciences of the University of Paris in 1895. In it he stated precisely in the following few words the object and results of his work:

"From the point of view of their magnetic properties, bodies may be divided into two groups: diamagnetic bodies, bodies only feebly magnetic, and paramagnetic bodies.[4] At first sight the two groups seem entirely separate. The principal aim of this research has been to discover if there exist transitions between these two states of matter, and if it is possible to make a given body pass progressively through them. To determine this I have examined the properties of a great number of bodies at temperatures differing as much as possible, in magnetic fields of varying intensities.

"My experiments failed to prove any relation between the properties of diamagnetic and those of paramagnetic bodies. And the results support the theories which attribute magnetism and diamagnetism to causes of a different nature. On the contrary, the properties of ferro-magnetic bodies and of bodies feebly magnetic are intimately united."

This experimental work presented many difficulties, for it necessitated the measuring of very minute forces (of the order of 1/100 of a milligramme weight) within a container where the temperature could attain 400° C.

As Pierre Curie well understood, the results he obtained are, from a theoretic point of view, of fundamental importance. The Curie law, according to which the coefficient of magnetization of a body feebly magnetized varies in inverse ratio to the absolute temperature, is a remarkably simple law. It is quite comparable to the Gay-Lussac law relating to the variation of the density of a perfect gas with the temperature. In his well known theory of magnetism P. Langevin, in 1905, took into account the Curie law and arrived again, theoretically, at the difference between the origins of diamagnetism and paramagnetism. His work, as well as the important investigations of P. Weiss, demonstrated the accuracy of Pierre Curie's conclusions, as well as the importance of the analogy that he perceived between the intensity of magnetization and the density of a fluid—the paramagnetic state being comparable to a gaseous state, and the ferro-magnetic state to the state of condensation.

In connection with this work, Pierre Curie spent some time in the search for unknown phenomena whose existence did not seem, a priori, impossible to him. He sought for bodies strongly diamagnetic, but found none. He tried to discover, too, if there were bodies that acted as conductors of magnetism, and if magnetism can exist in a "free state," like electricity. Here also the result was negative. He never published any of these investigations, for he had the habit of thus engaging in the pursuit of phenomena, often with little hope of success, solely for the love of the unforeseen, and without ever thinking of publication.

Because of this entirely disinterested passion for scientific research the presentation of a doctor's thesis which would give an account of these early investigations had never appealed to him. He was already thirty-five years old when he decided to gather together, in such a thesis, the results of his beautiful work on magnetism.

I have a very vivid memory of how he sustained his thesis before the examiners, for he had invited me, because of the friendship that already existed between us, to be present on the occasion. The jury was composed of Professors Bonty, Lippmann, and Hautefeuille. In the audience were some of his friends, among them his aged father, extremely happy in his son's success. I remember the simplicity and the clarity of the exposition, the esteem indicated by the attitude of the professors, and the conversation between them and the candidate which reminded one of a meeting of the Physics Society. I was greatly impressed; it seemed to me that the little room that day sheltered the exaltation of human thought.

In recalling this period in the life of Pierre Curie, between 1883 and 1895, we can appreciate the great progress the young physicist had made while acting as Chief of Laboratory. He had succeeded during this time in organizing an entirely new teaching service; he had published an important series of theoretical memoirs, as well as the results of experimental research of the first order. In addition, he had constructed new apparatus of great perfection—and all this in spite of very insufficient accommodations and resources. This achievement suggests the distance he had traveled since the doubts and hesitations of his early youth in learning to discipline his methods of work, and to derive from them the full advantage of his exceptional capacities.

He enjoyed a growing esteem in France, and in foreign countries. He was listened to with interest at the meetings of the learned societies (Society of Physics, Society of Mineralogy, Society of Electricians), where he was in the habit of presenting his communications and where he joined readily in the discussion of various scientific questions.

Among foreign scholars who already at this time appreciated him highly, I can name, in the first place, the illustrious English physicist, Lord Kelvin, who joined with him in a certain scientific discussion, and who often expressed for him, from that time on, both esteem and sympathy. During one of his visits to Paris, Lord Kelvin was present at a meeting of the Society of Physics when Pierre Curie made a statement regarding the construction and the use of standard condensers with guard ring. In this statement he recommended the use of an apparatus which involved the charging of the central part of the guard ring plate by a galvanic cell and in uniting the guard ring with the earth. One uses then, as a measure, the charge induced on the second plate. Even though the resulting disposition of lines of the field be complex, the charge induced can be calculated by a theorem of electrostatics, with the same simple formula as is used for an ordinary apparatus in a uniform field, and one has the benefit of a better isolation. Lord Kelvin believed at first that this reasoning was inexact. Despite his great repute and his advanced age, he went the following day to the laboratory to find the young Director. Here he discussed the matter with him before the blackboard. He was completely convinced, and seemed even delighted to concede the point to his companion.[5]

It may seem astonishing that Pierre Curie, in spite of his merits, continued during twelve years in the small position of Chief of Laboratory. Without doubt this was largely due to the fact that it is easy to overlook those who have not the active support of influential persons. It was due also to the fact that it was impossible for him to take the many steps that the pushing of any candidature involves. Then, too, his independence of character ill fitted him to ask for an advance, and this notwithstanding the fact that his position was very modest. Indeed his salary, then comparable to that of a day laborer (about 300 francs a month), was scarcely sufficient to enable him to lead the simple life that would yet permit him to carry on his work.

He expressed his feelings on this subject in the following words:

"I have heard that perhaps one of the professors will resign, and that I might, in that case, make application to succeed him. What an ugly necessity is this of seeking any position whatsoever; I am not accustomed to this form of activity, demoralizing to the highest degree. I am sorry that I spoke to you about it. I think that nothing is more unhealthy to the spirit than to allow oneself to be occupied by things of this character and to listen to the petty gossip that people come to report to you."

If he disliked soliciting an advancement in position, he was even less inclined to hope for honors. He had, in fact, a very decided opinion on the subject of honorary distinctions. Not only did he believe that they were not helpful, but he considered them frankly harmful. He felt that the desire to obtain them is a cause of trouble, and that it can degrade the worthiest aim of man, which is, work for the pure love of it.

Since he possessed great moral probity, he did not hesitate to make his acts conform to his opinions. When Schützenberger, in order to offer him a mark of esteem, wished to propose him for the Palmes académiques he refused this distinction, despite the advantages which, according to general belief, it would confer. And he wrote to his director:

"I have been informed that you intend to propose me again to the prefet for the decoration. I pray you do not do so. If you procure for me this honor, you will place me under the necessity of refusing it, for I have firmly decided not to accept a decoration of any kind. I hope that you will be good enough to avoid taking a step that will make me appear a little ridiculous in the eyes of many people. If your aim is to offer me a testimony of your interest, you have already done that, and in a very much more effective manner which touched me greatly, for you have made it possible for me to work without worry."

Faithful to this firm opinion, he later declined the decoration of the Légion d'Honneur, which was offered him in 1903.

But even though Pierre Curie refused to take steps to change his situation it was at last improved. In 1895 the well-known physicist, Mascart, professor in the Collège de France, impressed with his ability, and with Lord Kelvin's opinion of him, insisted that Schützenberger create a new Chair of Physics at the School of Physics and Chemistry. Pierre Curie was then named professor under conditions in which his talents were duly recognized. However, nothing was done at this time to ameliorate the inadequate material conditions under which, as we have already seen, he was carrying on his personal investigations.

[3]In this very brief memoir is presented, for the first time, a theory which explains why crystals develop certain faces simultaneously, in a particular direction, and consequently why crystals possess a determined form.

[4]Paramagnetic bodies are those which are magnetized in the same manner as iron, either strongly (ferro-magnetic) or feebly. Diamagnetic bodies are those whose very feeble magnetization is opposed to that which iron takes in the same magnetic field.

[5]The following is the text of a letter from this distinguished savant to Pierre Curie, written during one of his visits to Paris:

October, 1893.

"DEAR MR. CURIE:

"I am much obliged to you for your letter of Saturday and the information contained in it, which is exceedingly interesting to me.

"If I call at your laboratory between 10 and 11 tomorrow morning should I find you there? There are two or three things I would like to speak to you about; and I would like also to see more of your curves representing the magnetization of iron at different temperatures.

"Yours truly,

"KELVIN."

# CHAPTER IV

## MARRIAGE AND ORGANIZATION OF THE FAMILY LIFE. PERSONALITY AND CHARACTER

I met Pierre Curie for the first time in the spring of the year 1894. I was then living in Paris where for three years I[6] had been studying at the Sorbonne. I had passed the examinations for the licentiate in physics, and was preparing for those in mathematics. At the same time I had begun to work in the research laboratory of Professor Lippmann. A Polish physicist whom I knew, and who was a great admirer of Pierre Curie, one day invited us together to spend the evening with himself and his wife.

As I entered the room, Pierre Curie was standing in the recess of a French window opening on a balcony. He seemed to me very young, though he was at that time thirty-five years old. I was struck by the open expression of his face and by the slight suggestion of detachment in his whole attitude. His speech, rather slow and deliberate, his simplicity, and his smile, at once grave and youthful, inspired confidence. We began a conversation which soon became friendly. It first concerned certain scientific matters about which I was very glad to be able to ask his opinion. Then we discussed certain social and humanitarian subjects which interested us both. There was, between his conceptions and mine, despite the difference between our native countries, a surprising kinship, no doubt attributable to a certain likeness in the moral atmosphere in which we were both raised by our families.

We met again at the Physics Society and in the laboratory. Then he asked if he might call upon me. I lived at that time in a room on the sixth floor of a house situated near the schools. It was a poor little room, for my resources were extremely limited. I was, nevertheless, very happy in it for I was now first realizing, although already twenty-five years old, the ardent desire I had so long cherished of carrying on advanced studies in science.

Pierre Curie came to see me, and showed a simple and sincere sympathy with my student life. Soon he caught the habit of speaking to me of his dream of an existence consecrated entirely to scientific research, and he asked me to share that life. It was not, however, easy for me to make such a decision, for it meant separation from my country and my family, and the renouncement of certain social projects that were dear to me. Having grown up in an atmosphere of patriotism kept alive by the oppression of Poland, I wished, like many other young people of my country, to contribute my effort toward the conservation of our national spirit.

So matters stood, when at the beginning of my vacation I left Paris to go to my father in Poland. Our correspondence during this separation helped to strengthen the bond of affection between us.

During the year 1894 Pierre Curie wrote me letters that seem to me admirable in their form. No one of them was very long, for he had the habit of concise expression, but all were written in a spirit of sincerity and with an evident anxiety to make the one he desired as a companion know him as he was. The very quality of the expression has always seemed to me remarkable. No other one could describe in a few lines, as he could, a state of mind, or a situation, and by the simplest means make that description evoke a seizing image of truth. Because of this gift, he might, I believe, have been a great writer. I have already cited a few fragments of his letters, and others will follow. It is appropriate to quote here a few lines which express how he looked on the possibility of our marriage:

"We have promised each other (is it not true?) to have, the one for the other, at least a great affection. Provided that you do not change your mind! For there are no promises which hold; these are things that do not admit of compulsion.

"It would, nevertheless, be a beautiful thing in which I hardly dare believe, to pass through life together hypnotized in our dreams: your dream for your country; our dream for humanity; our dream for science. Of all these dreams, I believe the last, alone, is legitimate. I mean to say by this that we are powerless to change the social order. Even if this were not true we should not know what to do. And in working without understanding we should never be sure that we were not doing more harm than good, by retarding some inevitable evolution. From the point of view of science, on the contrary, we can pretend to accomplish something. The territory here is more solid and obvious, and however small it is, it is truly in our possession.

"I strongly advise you to return to Paris in October. I shall be very unhappy if you do not come this year, but it is not my friend's selfishness that makes me ask you to return. I ask it because I believe you will work better here and that you can accomplish here something more substantial and more useful."

One can understand, from this letter, that for Pierre Curie there was only one way of looking at the future. He had dedicated his life to his dream of science: he felt the need of a companion who could live his dream with him. He told me many times that the reason he had not married until he was thirty-six was because he did not believe in the possibility of a marriage which would meet this, his absolute necessity.

When he was twenty-two years old he wrote in his diary:

"Women, much more than men, love life for life's sake. Women of genius are rare. And when, pushed by some mystic love, we wish to enter into a life opposed to nature, when we give all our thoughts to some work which removes us from those immediately about us, it is with women that we have to struggle, and the struggle is nearly always an unequal one. For in the name of life and of nature they seek to lead us back."

We can see, however, in the letters I have quoted earlier, the unshakeable faith that Pierre Curie had in science and in its power to further the general good of humanity. It seems appropriate to apply to him the sentiment expressed by Pasteur in words so well known: "I believe invincibly that science and peace will triumph over ignorance and war."

This confidence in the solutions of science made Pierre Curie little inclined to take an active part in politics. He was attached, by education and by conviction, to democratic and socialistic ideas, but he was not dominated by any party doctrine. However, he always fulfilled, as his father did, his obligations as a voter. In public life, as in private life, he was opposed to the use of violence.

"What would you think," he wrote me, "of a person who would knock his head against a stone wall with the intention of overthrowing it? Such an idea might be the result of very beautiful feelings, but in realization it would be ridiculous and stupid. I believe that certain questions demand a general solution, and do not admit, today, of specific solutions, and that one who begins a course that has no issue, may do much harm. I believe, further, that justice is not of this world, and that the strongest system or rather the one best developed from the economic point of view will be that which will stand. A man may exhaust himself by work, and yet live, at best, miserably. This is a revolting fact, but it will not, because of that, cease. It will disappear probably because man is a kind of machine, and it is of economic advantage to make every machine work in its normal manner, without forcing it."

He felt the same necessity for clarity and understanding in considering his own inner life as in examining a general problem. A great necessity of loyalty to himself and toward others made him suffer from the compromises imposed by life, even though he reduced them to a minimum.

"We are all the slaves of our affections," he wrote, "slaves of the prejudices of those we love. Besides, we must make a living, and this forces us to become a wheel in the machine. The most painful are the concessions we are forced to make to the prejudices of the society in which we live. We must make more or fewer compromises according as we feel ourselves feebler or stronger. If one does not make enough concessions he is crushed; if he makes too many he is ignoble and despises himself. I find myself far from the principles I held ten years ago. At that time I believed it necessary to be excessive in everything, and to make no concessions whatsoever to one's environment. I believed it necessary to exaggerate one's faults as well as one's virtues."

This was the credo of the man who, without fortune himself, desired to share his life with that of a student also without fortune, whom he had met by chance.

After my return from my vacation our friendship grew more and more precious to us; each realized that he or she could find no better life companion. We decided, therefore, to marry, and the ceremony took place in July, 1895. In conformity with our mutual wish it was the simplest service possible,—a civil ceremony, for Pierre Curie professed no religion, and I myself did not practice any. My husband's parents received me with great cordiality, and reciprocally my father and my sisters, who were present at our marriage, were happy in knowing the family to which I was to belong.

Our first home, an extremely simple one, consisted of a little apartment of three rooms in the rue de la Glacière, not far from the School of Physics. Its chief attraction was its view of a large garden. It was furnished very simply with objects that had belonged to our families. Our means did not permit our having servants, so that I had to assume practically all the household duties, as I had been in the habit of doing during my student days.

Professor Curie's salary was 6000 francs a year, and we held that he should not undertake any supplementary work, at least in the beginning. As for myself, I was preparing to take the examination for the agregation of young women, in view of obtaining a teaching post. These I passed in 1896. We ordered our life to suit our scientific work and our days were passed in the laboratory, where Schützenberger permitted that I might work with my husband.

He was then engaged in a research on the growth of crystals, which interested him keenly. He wished to know if certain faces of a crystal had a preferential development chiefly because they have a different rapidity of growth or because their solubility is different. He quickly obtained interesting results (not published) but he had to interrupt his investigations to undertake others on radioactivity. And he often regretted that he was never able to return to them. I was occupied at this time with the study of the magnetization of tempered steel.

The preparation of his class lectures was for Pierre Curie a genuine care. The Chair was a new one, and carried no prescribed course of study. He divided his lectures, at first, between crystallography and electricity. Then, as he recognized more and more the utility of a serious theoretical course in electricity for future engineers, he devoted himself entirely to this subject, and succeeded in establishing a course (of about 120 lectures) that was the most complete and modern then to be had in Paris. This cost him a considerable effort, of which I was the daily witness; for he was always anxious to give a complete picture of the phenomena and of the evolution of theories, and of ideas. He was always anxious, too, that his mode of exposition should be clear and precise. He thought of publishing a treatise summing up this course, but unfortunately the many preoccupations of the following years prevented him from putting this plan into execution.

We lived a very single life, interested in common, as we were, in our laboratory experiments and in the preparation of lectures and examinations. During eleven years we were scarcely ever separated, which means that there are very few lines of existing correspondence between us, representing that period. We spent our rest days and our vacations walking or bicycling either in the country near Paris, or along the sea, or in the mountains. My husband was so engrossed in his researches, however, that it was very difficult for him to remain for any length of time in a place where he lacked facilities for work. After a few days he would say: "It seems to me a very long time since we have accomplished anything." And yet he liked the excursions which covered successive days, and enjoyed to the full our walks together, just as he had formerly enjoyed those with his brother. But his joy in seeing beautiful things never drew his thoughts away from the scientific questions that absorbed him. In these free times, we traversed the region of the Cevennes and of the Monts d'Auvergne, as well as the coast of France, and some of its great forests.

These days in the open, filled with beautiful sights, made a deep impression on us, and we loved to recall them. One of our radiant memories was of a sunny day, when after a long and wearying climb, we reached the fresh, green meadow of the Aubrac, in the pure air of the high plateaus. Another vivid memory was that of an evening, when, lingering until twilight in the gorge of the Truyère, we were enchanted to hear a popular air dying away in the distance, carried to us from a little boat that descended the stream. We had taken so little notice of the time that we did not regain our lodging before dawn. At one point we had an encounter with carts whose horses were frightened by our bicycles, and we were obliged to cut across ploughed fields. At length we regained our route on the high plateau, bathed in the unreal light of the moon. And cows that were passing the night in enclosures came gravely to contemplate us with their large, tranquil eyes.

The forest of Compiegne charmed us in the spring, with its mass of green foliage stretching far as the eye could see, and its periwinkles and anemones. On the border of the forest of Fontainebleau, the banks of the Loing, covered with water buttercups, were an object of delight for Pierre Curie. We loved the melancholy coasts of Brittany and the reaches of heather and gorse, stretching to the very points of Finistère, which seemed like claws or teeth burying themselves in the water which forever rages at them.

Later, when we had our baby with us, we passed our vacations in some one locality, without traveling about. We lived then as simply as possible in retired villages where we could scarcely be distinguished from the villagers themselves. I remember the stupefaction of an American journalist when he found us one day at Pouldu, at a moment when I was sitting on one of the stone steps of our house in the act of shaking the sand from my sandals. However, his embarrassment was short-lived and, adapting himself to the situation, he sat down beside me and began jotting down in his notebook my answers to his questions.

The most affectionate relations existed between my husband's parents and myself. We often went to Sceaux, where the room my husband used to have before our marriage was always reserved for us. I had also a very tender affection for Jacques Curie and his family (he was married and had two children); for Pierre's brother became mine, and has always remained so.

Our eldest daughter, Irene, was born in September, 1897, and only a few days afterwards my husband suffered a great loss in the death of his mother. Doctor Curie came to live with us in a house which had a garden and was situated on the old fortifications of Paris (108 Boulevard Kellermann) near the park of Montsouris. Pierre Curie lived in this house until the end of his life.

With the birth of our child, the difficulties of carrying on our work were augmented: for I had to give more time to the household. Very fortunately for us I could leave my little girl with her grandfather, who much enjoyed taking care of her. But we had to think also of increasing our resources to meet the needs of our larger family and to enable us to secure someone to help me in the house, a necessity from now on. However, our situation remained as it was during the following two years, which we consecrated to intensive laboratory research on radioactivity. It was, indeed, not relieved until 1900, to the detriment, it is true, of the amount of time we could give to our investigations.

All formal social obligations were excluded from our life. Pierre Curie had for such things an unconquerable repugnance. Neither in his earlier nor his later life would he pay visits or undertake to involve himself in relations without special interest. By nature grave and silent, he preferred to abandon himself to his own reflections, rather than to engage in an exchange of banal words. On the other hand, he valued greatly his boyhood friends, and those to whom he was bound by a common interest in science.

Among the latter, E. Gouy, professor of the faculty of sciences at Lyon, should be named. His friendly relations with Pierre Curie dated from the time when they were both preparators at the Sorbonne. They carried on regularly a scientific correspondence, and took great pleasure in seeing each other again during the various brief visits of E. Gouy to Paris, on which occasions they were inseparable. There existed also a friendship of long standing between my husband and Ch. Ed. Guillaume, now director of the International Bureau of Weights and Measures of Sèvres. They met at the Physics Society and occasionally on Sundays at Sèvres or Sceaux. Later a group of younger men formed themselves about Pierre Curie. They were investigators engaged, as he was, in physical and chemical research in the newest fields of these sciences. Among these men were André Debierne, my husband's intimate friend and collaborator in the work on radioactivity; George Sagnac, his collaborator in a study of the X-rays; Paul Langevin, who became a professor in the Collège de France; Jean Perrin, at present professor of physical chemistry in the Sorbonne; and Georges Urbain, student of the School of Physics and later professor in the Sorbonne. Often one or the other came to see us in our quiet house in the Boulevard Kellermann. Then we engaged in discussions of recent or future experiments, or of new ideas and theories, and never tired of rejoicing over the marvelous development of modern physics.

There were not many large reunions in our house, for my husband did not feel the need of them. He was more at his ease in a conversation with some one or few persons, and rarely attended any meetings except those of the scientific societies. If by chance he found himself in a gathering where the general conversation did not interest him, he took refuge in a tranquil corner where he could forget the company as he pursued his own thoughts.

Our relations with our families were very restricted on his side as on mine; for he had few relatives and mine were far away. He was, however, very devoted to those of my family who could come to visit me in Paris, or during our vacations.

In 1899, Pierre Curie made a journey with me to the Carpathians of Austrian Poland, where one of my sisters, married to Doctor Dluski and herself a physician, directed, with him, a large sanatorium. Through a touching desire to know all that was dear to me, my husband, though he knew little of foreign languages, wished to learn Polish, something which I had not thought of suggesting because I did not believe it could prove sufficiently useful to him. He felt a sincere sympathy for my country and believed in the future Establishment of a free Poland.

In our life together it was given to me to know him as he had hoped I might, and to penetrate each day further into his thought. He was as much and much more than all I had dreamed at the time of our union. My admiration of his unusual qualities grew continually; he lived on a plane so rare and so elevated that he sometimes seemed to me a being unique in his freedom from all vanity and from the littlenesses that one discovers in oneself and in others, and which one judges with indulgence although aspiring to a more perfect ideal.

In this lay, without doubt, the secret of that infinite charm of his to which one could not long rest insensible. His thoughtful expression and the directness of his look were strongly attractive and this attraction was increased by his kindliness and gentleness of character. He sometimes said that he never felt combative, and this was entirely true. One could not enter into a dispute with him because he could not become angry. "Getting angry is not one of my strong points," he would say, smiling. If he had few friends, he had no enemies; for he could not injure anyone, even by inadvertence. But at the same time no one could force him to deviate from his line of action, something which led his father to nickname him the "gentle stubborn one."

When he expressed his opinion he did so frankly, for he was convinced that diplomatic methods are puerile, and that directness is at once easiest and best. Because of this practice, he acquired a certain reputation for naïveté; in reality he was acting on a well-considered decision, rather than by instinct. It was perhaps because he was able to judge himself and to retire within himself, that he was so capable of clearly appreciating the springs of action, the intention, and the thoughts of others. And if he sometimes neglected details, he was rarely deceived in the essentials. Usually he kept his sure judgments to himself; but once he had made up his mind he sometimes expressed them without reticence, in the assurance that he was doing something useful.

In his scientific relations he showed no sharpness, and did not permit himself to be influenced by considerations of personal credit or by personal sentiments. Every beautiful success gave him pleasure, even if achieved in a domain where he felt himself to have priority. He said: "What does it matter if I have not published such and such investigations, if another has published them?" For he held that in science we should be interested in things and not in persons. He was so genuinely against every form of emulation that he opposed even the competitions and gradings of the lycées, as well as all forms of honorary distinction. He never failed to give counsel and encouragement to any of those who showed an aptitude for science, and certain among them still remain profoundly grateful to him.

If his attitude was that of one of the élite who have attained the highest summit of civilization, his acts were those of a truly good man endowed with the sentiment of human solidarity intimately bound to his intellectual conceptions, and full of understanding and indulgence. He was always ready to aid, as far as his means allowed, any person in a difficult situation, even if helping meant giving some of his time, which was always the greatest sacrifice he could make. His generosity was so spontaneous that one scarcely noticed it. He believed that the only advantage of material means, beyond that of providing the necessities of a simple life, was in the opportunity they offered of aiding others, and of pursuing the work of one's preference.

What shall I say, finally, of his love for his own, and of his qualities as friend? His friendship, which he gave rarely, was sure and faithful, for it rested on a community of ideas and opinions. And still more rarely did he give affection; but how complete was his gift to his brother and to me! He could forsake his customary reserve for an unconstraint which established harmony and confidence. His tenderness was the most exquisite of blessings, sure and helpful, full of gentleness and solicitude. It was good to be surrounded by this tenderness; it was cruel to lose it after having lived in an atmosphere completely permeated by it. But I will let his own words tell how completely he gave himself:

"I think of you who fill my life, and I long for new powers. It seems to me that in concentrating my mind exclusively upon you, as I am doing, that I should succeed in seeing you, and in following what you are doing; and that I should be able to make you feel that I am altogether yours at this moment,—but the image does not come."

We were not warranted in having great confidence in our health, nor in our strength so often put to severe tests. And from time to time, as happens to those who know the value of sharing a common life, the fear of the irreparable touched our minds. In such moments his simple courage led him always to the same inevitable conclusion: "Whatever happens, even if one should become like a body without a soul, still one must always work."

[6]The following are a few brief biographical details:

My name is Marie Sklodowska. My father and mother belonged to Catholic Polish families. Both were teachers in secondary schools in Warsaw (at that time under Russia). I was born in Warsaw and attended a lycée there. Following the lycée, I taught several years. Then in 1892 I came to Paris in order to study science.

# CHAPTER V

## THE DREAM BECOME A REALITY. THE DISCOVERY OF RADIUM

I have already said that in 1897 Pierre Curie was occupied with an investigation on the growth of crystals. I myself had finished, by the beginning of vacation, a study of the magnetization of tempered steels which had resulted in our getting a small subvention from the Society for the Encouragement of National Industry. Our daughter Irene was born in September, and as soon as I was well again, I resumed my work in the laboratory with the intention of preparing a doctor's thesis.

Our attention was caught by a curious phenomenon discovered in 1896 by Henri Becquerel. The discovery of the X-ray by Roentgen had excited the imagination, and many physicians were trying to discover if similar rays were not emitted by fluorescent bodies under the action of light. With this question in mind Henri Becquerel was studying uranium salts, and, as sometimes occurs, came upon a phenomenon different from that he was looking for: the spontaneous emission by uranium salts of rays of a peculiar character. This was the discovery of radioactivity.

The particular phenomenon discovered by Becquerel was as follows: uranium compound placed upon a photographic plate covered with black paper produces on that plate an impression analogous to that which light would make. The impression is due to uranium rays that traverse the paper. These same rays can, like X-rays, discharge an electroscope, by making the air which surrounds it a conductor.

Henri Becquerel assured himself that these properties do not depend on a preliminary isolation, and that they persist when the uranium compound is kept in darkness during several months. The next step was to ask whence came this energy, of minute quantity, it is true, but constantly given off by uranium compounds under the form of radiations.

The study of this phenomenon seemed to us very attractive and all the more so because the question was entirely new and nothing yet had been written upon it. I decided to undertake an investigation of it.

It was necessary to find a place in which to conduct the experiments. My husband obtained from the director of the School the authorization to use a glassed-in study on the ground floor which was then being used as a storeroom and machine shop.

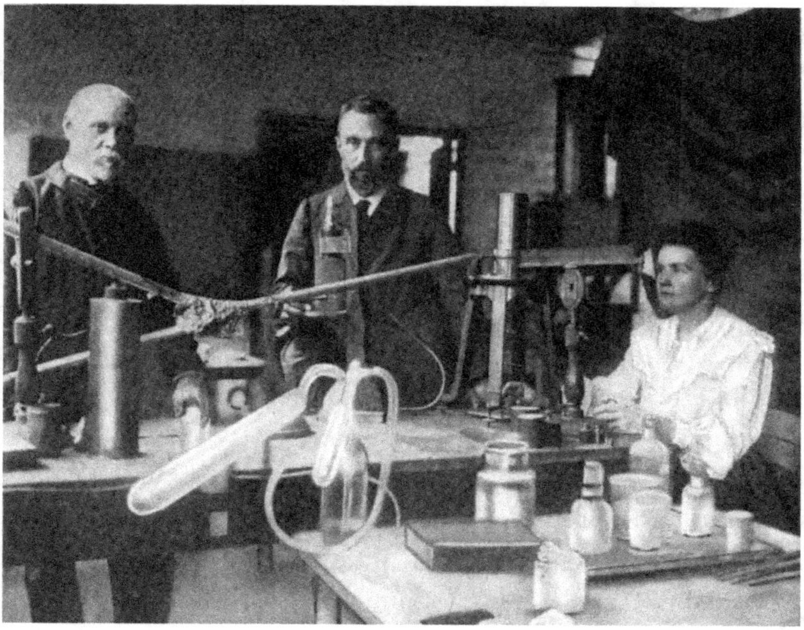

Pierre and Marie Curie in their laboratory, where radium was discovered.

Henri Manuel, Paris.

In order to go beyond the results reached by Becquerel, it was necessary to employ a precise quantitative method. The phenomenon that best lent itself to measurement was the conductibility produced in the air by uranium rays. This phenomenon, which is called ionization, is produced also by X-rays and investigation of it in connection with them had made known its principal characteristics.

For measuring the very feeble currents that one can make pass through air ionized by uranium rays, I had at my disposition an excellent method developed and applied by Pierre and Jacques Curie. This method consists in counterbalancing on a sensitive electrometer the quantity of electricity carried by the current with that which a piezo-electric quartz can furnish. The installation therefore required a Curie electrometer, a piezo-electric quartz, and a chamber of ionization, which last was formed by a plate condenser whose higher plate was joined to the electrometer, while the lower plate, charged with a known potential, was covered with a thin layer of the substance to be examined. Needless to say, the place for such an electrometric installation was hardly the crowded and damp little room in which I had to set it up.

My experiments proved that the radiation of uranium compounds can be measured with precision under determined conditions, and that this radiation is an atomic property of the element of uranium. Its intensity is proportional to the quantity of uranium contained in the compound, and depends neither on conditions of chemical combination, nor on external circumstances, such as light or temperature.

I undertook next to discover if there were other elements possessing the same property, and with this aim I examined all the elements then known, either in their pure state or in compounds. I found that among these bodies, thorium compounds are the only ones which emit rays similar to those of uranium. The radiation of thorium has an intensity of the same order as that of uranium, and is, as in the case of uranium, an atomic property of the element.

It was necessary at this point to find a new term to define this new property of matter manifested by the elements of uranium and thorium. I proposed the word radioactivity which has since become generally adopted; the radioactive elements have been called radio elements.

During the course of my research, I had had occasion to examine not only simple compounds, salts and oxides, but also a great number of minerals. Certain ones proved radioactive; these were those containing uranium and thorium; but their radioactivity seemed abnormal, for it was much greater than the amount I had found in uranium and thorium had led me to expect.

This abnormality greatly surprised us. When I had assured myself that it was not due to an error in the experiment, it became necessary to find an explanation. I then made the hypothesis that the ores uranium and thorium contain in small quantity a substance much more strongly radioactive than either uranium or thorium. This substance could not be one of the known elements, because these had already been examined; it must, therefore, be a new chemical element.

I had a passionate desire to verify this hypothesis as rapidly as possible. And Pierre Curie, keenly interested in the question, abandoned his work on crystals (provisionally, he thought) to join me in the search for this unknown substance.

We chose, for our work, the ore pitchblende, a uranium ore, which in its pure state is about four times more active than oxide of uranium.

Since the composition of this ore was known through very careful chemical analysis, we could expect to find, at a maximum, 1 per cent of new substance. The result of our experiment proved that there were in reality new radioactive elements in pitchblende, but that their proportion did not reach even a millionth per cent!

The method we employed is a new method in chemical research based on radioactivity. It consists in inducing separation by the ordinary means of chemical analysis, and of measuring, under suitable conditions, the radioactivity of all the separate products. By this means one can note the chemical character of the radioactive element sought for, for it will become concentrated in those products which will become more and more radioactive as the separation progresses. We soon recognized that the radioactivity was concentrated principally in two different chemical fractions, and we became able to recognize in pitchblende the presence of at least two new radioactive elements:

polonium and radium. We announced the existence of polonium in July, 1898, and of radium in December of the same year.[7]

In spite of this relatively rapid progress, our work was far from finished. In our opinion, there could be no doubt of the existence of these new elements, but to make chemists admit their existence, it was necessary to isolate them. Now, in our most strongly radioactive products (several hundred times more active than uranium), the polonium and radium were present only as traces. The polonium occurred associated with bismuth extracted from pitchblende, and radium accompanied the barium extracted from the same mineral. We already knew by what methods we might hope to separate polonium from bismuth and radium from barium; but to accomplish such a separation we had to have at our disposition much larger quantities of the primary ore than we had.

It was during this period of our research that we were extremely handicapped by inadequate conditions, by the lack of a proper place to work in, by the lack of money and of personnel.

Pitchblende was an expensive mineral, and we could not afford to buy a sufficient quantity. At that time the principal source of this mineral was at St. Joachimsthal (Bohemia) where there was a mine which the Austrian government worked for the extraction of uranium. We believed that we would find all the radium and a part of the polonium in the residues of this mine, residues which had so far not been used at all. Thanks to the influence of the Academy of Sciences of Vienna, we secured several tons of these residues at an advantageous price, and we used it as our primary material. In the beginning we had to draw on our private resources to pay the costs of our experiment; later we were given a few subventions and some help from outside sources.

The question of quarters was particularly serious; we did not know where we could conduct our chemical treatments. We had been obliged to start them in an abandoned storeroom across a court from the workroom where we had our electrometric installation. This was a wooden shed with a bituminous floor and a glass roof which did not keep the rain out, and without any interior arrangements. The only objects it contained were some worn pine tables, a cast-iron stove, which worked badly, and the blackboard which Pierre Curie loved to use. There were no hoods to carry away the poisonous gases thrown off in our chemical treatments, so that it was necessary to carry them on outside in the court, but when the weather was unfavorable we went on with them inside, leaving the windows open.

A view of the extraction of radium in the old shed where the first radium was obtained.

In this makeshift laboratory we worked practically unaided during two years, occupying ourselves as much with chemical research as with the study of the radiation of the increasingly active products we were obtaining. Then it became necessary for us to divide our work. Pierre Curie continued the investigations on the properties of radium, while I went ahead with the chemical experiments which had as their objective the preparation of pure radium salts. I had to work with as much as twenty kilogrammes of material at a time, so that the hangar was filled with great vessels full of precipitates and of liquids. It was exhausting work to move the containers about, to transfer the liquids, and to stir for hours at a time, with an iron bar, the boiling material in the cast-iron basin. I extracted from the mineral the radium-bearing barium and this, in the state of chloride, I submitted to a fractional crystallization. The radium accumulated in the least soluble parts, and I believed that this process must lead to the separation of the chloride of radium. The very delicate operations of the last crystallizations were exceedingly difficult to carry out in that laboratory, where it was impossible to find protection from the iron and coal dust. At the end of a year, results indicated clearly that it would be easier to separate radium than polonium; that is why we concentrated our efforts in this direction. We examined the radium salts we obtained with the aim of discovering their powers and we loaned samples of the salts to several scientists,[8] in particular to Henri Becquerel.

During the years 1899 and 1900, Pierre Curie published with me a memoir on the discovery of the induced radioactivity produced by radium. We published another paper on the effects of the rays: the luminous effects, the chemical effects, etc.; and still another on the electric charge carried by certain of the rays. And, finally, we made a general report on the new radioactive substances and their radiations, for the Congress of Physics which met in Paris in 1900. My husband published, besides, a study of the action of a magnetic field on radium rays.

Pierre Curie with the quartz piezo-electroscope he invented, by which rays of radium are measured.

The main result of our investigations and of those of other scientists during these years, was to make known the nature of the rays emitted by radium, and to prove that they belonged to three different categories. Radium emits a stream of active corpuscles moving with great speed. Certain of them carry a positive charge and form the Alpha rays; others, much smaller, carry a negative charge and form

Beta rays. The movements of these two groups are influenced by a magnet. A third group is constituted by the rays that are insensible to the action of a magnet, and that, we know to-day, are a radiation similar to light and to X-rays.

We had an especial joy in observing that our products containing concentrated radium were all spontaneously luminous. My husband who had hoped to see them show beautiful colorations had to agree that this other unhoped-for characteristic gave him even a greater satisfaction than that he had aspired to.

The Congress of 1900 offered us an opportunity to make known, at closer range, to foreign scientists, our new radioactive bodies. This was one of the points on which the interest of this Congress chiefly centered.

We were at this time entirely absorbed in the new field that opened before us, thanks to the discovery so little expected. And we were very happy in spite of the difficult conditions under which we worked. We passed our days at the laboratory, often eating a simple student's lunch there. A great tranquillity reigned in our poor, shabby hangar; occasionally, while observing an operation, we would walk up and down talking of our work, present and future. When we were cold, a cup of hot tea, drunk beside the stove, cheered us. We lived in a preoccupation as complete as that of a dream.

Sometimes we returned in the evening after dinner for another survey of our domain. Our precious products, for which we had no shelter, were arranged on tables and boards; from all sides we could see their slightly luminous silhouettes, and these gleamings, which seemed suspended in the darkness, stirred us with ever new emotion and enchantment.

Actually, the employees of the School owed Pierre Curie no service. But nevertheless the laboratory helper whom he had had to aid him when he was laboratory chief had always continued to help him as much as he could in the time at his disposal. This good man, whose name was Petit, felt a real affection and solicitude for us, and many things were made easier because of his good will and the interest he took in our success.

A view of the extraction of radium in the old shed where the first radium was obtained.

We had begun our research in radioactivity quite alone, but because of the magnitude of the undertaking, we were more and more convinced of the utility of inviting collaboration. Already in 1898, one of the laboratory chiefs of the School, G. Bemont, had given us temporary aid. And toward 1900 Pierre Curie associated with him a young chemist, André Debierne, preparator under Friedel, who held him in high esteem. André Debierne gladly accepted Pierre Curie's proposal that he occupy himself with the investigation of radioactivity; and he undertook, in particular, the search for a new radio element, which we suspected existed in the iron group and in rare earths. He discovered the element actinium. Even though he carried on his work in the laboratory of physical chemistry at the Sorbonne, directed by

Jean Perrin, he frequently came to visit us in our storeroom, and was soon an intimate friend of ours, and of Doctor Curie and the children.

About this same time, George Sagnac, a young physicist engaged in the study of X-rays, often came to discuss with my husband the analogies one could expect to find between these rays, and their secondary rays, and the radiations of radioactive bodies. They worked together on the investigation of the electric charge carried by the secondary rays.

Besides our collaborators we saw very few persons in the laboratory; however, from time to time some physicist or chemist came to see our experiments, or to ask Pierre Curie for advice or information; for his authority in several branches of physics was very well recognized. And then there were discussions before the blackboard,—discussions which are pleasantly remembered to-day, because they stimulated an interest in science and an ardor for work without interrupting any course of reflection, and without troubling that atmosphere of peace and contemplation which is the true atmosphere of the laboratory.

[7]This last publication was issued in common with G. Bemont, who had collaborated with us in our experiments.

[8]I quote, as an example, a letter addressed to Pierre Curie by A. Paulsen, thanking him for radioactive products loaned him in 1899:

"Den Damke Nordl's Expedition

Akureyi, 16 Oct. 1899.

Monsieur, and most honored colleague,

"I thank you warmly for your letter of August 1, which I have just received in the north of Iceland.

"We have abandoned all the methods hitherto employed to establish in a fixed conductor the potential that exists at certain points in the mass of air that surrounds it, and are using only your radiant powder.

"Accept, Monsieur, and most honored colleague, my respectful salutations and my renewed thanks for the great services you have rendered my expedition.

"ADAM PAULSEN."

# CHAPTER VI

## THE STRUGGLE FOR MEANS TO WORK. THE BURDEN OF CELEBRITY. THE FIRST ASSISTANCE FROM THE STATE. IT COMES TOO LATE

In spite of our desire to concentrate our entire effort on the work in which we were engaged, and in spite of the fact that our needs were so modest, we were forced to recognize, toward 1900, that some increase in our income was indispensable. Pierre Curie had few illusions about his chances of obtaining an important chair in the University of Paris, which would, even though it meant no large salary, have sufficed for the small needs of our family, and enabled us to live without a supplementary revenue. Since he was neither a graduate of the Normal School nor of the Polytechnic, he lacked the support, often decisive, which these big schools give their pupils; and the posts to which he might justly have aspired, because of his achievements, were given to others, without anyone's even thinking of him as a possible candidate. At the beginning of 1898, he asked, without success, for the Chair of Physical Chemistry left vacant by the death of Salet, and this failure convinced him that he had no chance of advancement. He was appointed, however, in March, 1900, to the position of assistant professor (répétiteur) in the Polytechnic School, but he kept his post only six months.

In the spring of 1900, there came an unexpected offer, that of the Chair of Physics in the University of Geneva. The doyen of that University made the invitation in the most cordial manner, and insisted that the University was ready to make an exceptional effort to secure a scientist of such high repute. The advantages of this position were that the salary was larger than the average one, that it carried the promise of the development of a Physics Laboratory adequate to our needs, and that an official position for me would be provided in this laboratory. Such a proposition merited a most careful consideration, so we made a visit to the University of Geneva, where our reception was the most encouraging possible.

This was a grave decision for us to make. Geneva presented material advantages, and the opportunity of a life comparable in its tranquillity with that in the country. Pierre Curie was, therefore, tempted to accept, and it was only our immediate interest in our researches in radium that made him finally decide not to. He feared the interruption of our investigations which such a change must involve.

At this moment the Chair of Physics in the physics, chemistry and natural history course at the Sorbonne, obligatory for students of medicine, and familiarly known as P.C.N., was vacant; he applied, and was appointed, due to the influence of Henri Poincaré, who was anxious to free him from the necessity of quitting France. At the same time I was given charge of the physics lectures in the Normal School for Girls at Sèvres.

So we remained in Paris, and with our income increased. But we were at the same time working under increasingly difficult conditions. Pierre Curie was doing double teaching; and that in the P.C.N., with its very large number of students, fatigued him greatly. As for myself, I had to give much time to the preparation of my lectures at Sèvres, and to the organization of the laboratory work there, which I found very insufficient.

Moreover, Pierre Curie's new position did not bring with it a laboratory; a little office and a single work room were all that he had at his disposition in the annex (12 rue Cuvier) of the Sorbonne, which served as teaching quarters for the P.C.N. And yet he felt it absolutely necessary to go ahead with his own work. In fact, the rapid extension of his investigations in radioactivity had made him determine that in his new position at the Sorbonne he would receive students and start them in research. He therefore took steps to find larger available working quarters. Those who have taken similar steps realize the wall of financial and administrative obstacles against which he was throwing himself, and realize the large number of official letters, visits, and of requests the least success entailed. All this thoroughly wearied and discouraged Pierre Curie. He was obliged, too, constantly, to keep traveling back and forth between the laboratories of the P.C.N. and the hangar of the School of Physics where we still continued our work.

And besides these difficulties, we found that we could not make further progress without the aid of industrial means of treating our raw material. Fortunately certain expedients and generous assistance solved this question.

As early as 1899 Pierre Curie succeeded in organizing a first industrial experiment, using for it a chance installation placed at his disposition by the Central Society of Chemical Products, with which he had had relations in connection with the construction of his balances. The technical details had been arranged very successfully by André Debierne, and the operations brought good results, even though it had been necessary to train a special personnel for this chemical work which demanded special precautions.

Our investigations had started a general scientific movement, and similar work was being undertaken in other countries. Toward these efforts Pierre Curie maintained a most disinterested and liberal attitude. With my agreement he refused to draw any material profit from our discovery. We took no copyright, and published without reserve all the results of our research, as well as the exact processes of the preparation of radium. In addition, we gave to those interested whatever information they asked of us. This was of great benefit to the radium industry, which could thus develop in full freedom, first in France, then in foreign countries, and furnish to scientists and to physicians the products which they needed. This industry still employs to-day, with scarcely any modifications, the processes indicated by us.[9]

Even though our industrial experiment yielded good results, again our slender resources made it difficult to make further progress. Inspired by our attempt, a French industrial, Armet de Lisle, had the idea, which seemed daring at that epoch, of founding a veritable radium factory that would furnish this product to physicians, whose interest in the biological effects of radium and its possible therapeutic applications had been aroused by the publication of various investigations. The project proved a success because he could employ men already trained by us in the delicate processes of this manufacture. Radium was then regularly placed on sale, at a high price, it is true, because of the special conditions under which it had to be made, and because, too, of the immediate rise in the cost of the minerals necessary to its production.[10]

I should like to express, here, our appreciation of the spirit in which Armet de Lisle offered to cooperate with us. In an entirely disinterested manner he placed at our disposition a little working place in his factory and a part of the means necessary for us to use it. Other funds were added either by ourselves, or came through subventions, of which the most important, accorded in 1902 by the Academy of Sciences, amounted to 20,000 francs.

It was in this way that we were able to utilize the ore we had acquired little by little in the preparation of a certain quantity of radium, which we used constantly in our research. The radium-bearing barium was extracted in the factory, and I carried on its purification and fractional crystallization in the laboratory. In 1902 I succeeded in preparing a decigramme of chloride of pure radium which gave only the spectrum of the new element, radium. I made a first determination of the atomic weight of this new element, an atomic weight much higher than that of barium. Thus the chemical individuality of radium was completely established, and the reality of radioelements was a known fact about which there could be no further controversy.

I based my doctor's thesis, presented in 1903, on these investigations.

Later, the quantity of radium extracted for the laboratory was increased, and in 1907 I was able to make a second and more precise determination of the atomic weight as 225.35—one accepts now the number 226. I succeeded, too, jointly with André Debierne, in obtaining radium in the state of metal. The total quantity of radium I prepared and gave to the laboratory, in agreement with Pierre Curie's desire, amounted to more than a gramme of radium element.

The activity of pure radium exceeded all our expectations. For equal weights this substance emits a radiation more than a million times more intense than uranium. To offset this, the quantity of radium contained in uranium minerals is scarcely more than three decigrammes of radium to the ton of uranium. There is a very close relation between these two substances. In fact, we know to-day that radium is produced in the minerals at the expense of uranium.

The years that followed his nomination to the P.C.N. were hard for Pierre Curie. He had to face the many anxieties incident to the organization of a complicated system of work when his happiness depended on his being able to concentrate his efforts on a single determined subject. The physical fatigue due to the numerous courses he was obliged to give was so great that he suffered from attacks of acute pain, which in his overtaxed condition became more and more frequent.

It was therefore vitally important, if he was to spare his energy and keep his health, that the burden of his professional duties be lightened. He decided to apply for the Chair of Mineralogy, which was vacant, at the Sorbonne, for which he was entirely qualified because of his profound knowledge and his important publications on the theories of the physics of crystals. Yet his candidacy failed.

During this painful period he nevertheless managed, by a truly superhuman effort, successfully to complete and publish several investigations that he had made either alone or in collaboration:

Investigations on induced radioactivity (in collaboration with A. Debierne).

Investigations on the same subject (in collaboration with J. Danne).

Investigations on the conductibility provoked in dielectric liquids by the rays of radium and the Roentgen rays.

Investigations on the law of the decrease of the emanation of radium and on the radioactive constants that characterize this emanation and its active deposit.

Discovery of the liberation of heat produced by radium (in collaboration with A. Laborde).

Investigations on the diffusion of the emanation of radium in the air (in collaboration with J. Danne).

Investigation on the radioactivity of gases from thermal springs (in collaboration with A. Laborde).

Investigation on the physiological effects of radium rays (in common with Henri Becquerel).

Investigation on the physiological action of the radium emanation (in common with Bouchard and Balthazard).

Notes on the apparatus for the determination of magnetic constants (in common with C. Cheneveau).

All these investigations in radioactivity are fundamental and touch very varied subjects. Several have as their aim the study of the emanation, that strange gaseous body that radium produces and which is largely responsible for the intense radiation commonly attributed to the radium itself. Pierre Curie demonstrated by a searching examination the rigorous and invariable law according to which the emanation destroys itself, no matter what the conditions are in which it finds itself. To-day the emanation of radium, harvested in tiny phials, is commonly employed by physicians as a therapeutic agent. Technical considerations make its employment preferable to the direct use of radium, and in this case no physician can proceed without consulting the numerical chart which tells how much of this emanation has disappeared each day, despite the fact that it is cloistered in its little glass prison. It is this same emanation that is found in small quantities in mineral waters, and that plays a part in their curative effects.

More striking still was the discovery of the discharge of heat from radium. Without any alteration in appearance this substance releases each hour a quantity of heat sufficient to melt its own weight of ice. When well protected against this external loss, radium heats itself. Its temperature can rise 10 degrees or more above that of the surrounding atmosphere. This defied all contemporary scientific experience.

Finally, I cannot pass in silence, because of their various repercussions, the experiments connected with the physiological effects of radium.

In order to test the results that had just been announced by F. Giesel, Pierre Curie voluntarily exposed his arm to the action of radium during several hours. This resulted in a lesion resembling a burn, that developed progressively and required several months to heal. Henri Becquerel had by accident a similar burn as a result of carrying in his vest pocket a glass tube containing radium salt. He came to tell us of this evil effect of radium, exclaiming in a manner at once delighted and annoyed: "I love it, but I owe it a grudge!"

Since he realized the interest in these physiological effects of radium, Pierre Curie undertook, in collaboration with physicians, the investigations to which I have just referred, submitting animals to the action of radium emanation. These studies formed the point of departure in radium therapy. The first attempts at treatment with radium were made with products loaned by Pierre Curie, and had as their object the cure of lupus and other skin lesions. Thus radium therapy, an important branch of medicine, and frequently designated as Curietherapie, was born in France, and was developed first through the investigations of French physicians (Danlos, Oudin, Wickham, Dominici, Cheron, Degrais, and others).[11]

In the meantime the great impetus given to the study of radioactivity abroad led to a rapid succession of new discoveries. Many scientists engaged in the search for other radio elements, using the new method of chemical analysis, with the aid of radiation, that we had inaugurated. Thus were found the mesothorium now used by physicians and manufactured industrially, radio-thorium, ionium, protoactinium, radio-lead, and other substances. At present we know, in all, about thirty radio elements (among which three are gases, or emanations), but among them all radium still plays the most important part, because of the great intensity of its radiation, which diminishes only extremely slowly during the course of years.

The year 1903 was especially important in the development of the new science. In this year the investigation of radium, the new chemical element, was achieved, and Pierre Curie demonstrated the astonishing discharge of heat by this element, which nevertheless remained unaltered in appearance. In England, Ramsay and Soddy announced a great discovery. They proved that radium continually produces helium gas and under conditions that force one to believe in an atomic transformation. If, indeed, radium salt heated to its melting point is confined for some time in a sealed glass tube, entirely emptied of air, one can, in reheating it, make it throw off a small quantity of helium, easy to measure and to recognize from the character of its spectrum. This fundamental experiment has received numerous confirmations. It furnished us the first example of a transformation of atoms, independent, it is true, of our will, but at the same time it reduces to nothing the theory of the absolute fixity of the atomic edifice.

All these facts, along with others formerly known, were made the object of a synthesis of the highest value, in a work by E. Rutherford and F. Soddy, who proposed a theory of radioactive transformations, to-day universally adopted. According to this theory, each radio element, even when it appears unchanged, is undergoing a spontaneous transformation, and the more rapid the transformation, the more intense is the radiation.[12]

A radioactive atom can transform itself in two ways: it can expel from itself an atom of helium, which, thrown off at an enormous speed and with a positive charge, constitutes an Alpha ray. Or, instead, it can detach from its structure a much smaller fragment, one of those electrons to which we have become accustomed in modern physics, and whose mass, 1800 times smaller than that of an atom of hydrogen when its speed is moderate, grows excessively when its speed approaches that of light. These electrons, which carry a negative charge, form the Beta rays. Whatever the detached fragment, the residual atom no longer resembles the primitive atom. Thus when the atom of radium has expelled an atom of helium, the residue is an atom of gaseous emanation. This residue changes in its turn, and the process is not arrested until the attainment of a last residue which is stable and does not give off any radiation. This stable matter is inactive matter.

Thus the Alpha and Beta rays result from the fragmentation of atoms. Gamma-rays are a radiation analogous to light, which accompanies the cataclysm of the atomic transformation. They are very penetrating, and are the ones most used in the therapeutic methods so far developed.[13]

We can see in all this that radio elements form families, in which each member derives from a preceding member by direct descent the primary elements being uranium and thorium. We can in particular prove that radium is a descendant of uranium, and that polonium is a descendant of radium. Since each radio element, at the same time that it is formed by the mother substance, destroys itself, it cannot accumulate in the presence of this mother substance beyond a determined proportion, which explains why the relation between radium and uranium remains constant in the very ancient unaltered minerals.

The spontaneous destruction of radio elements takes place according to a fundamental law, called the exponential law, according to which the quantity of each radio element diminishes by one-half in a time always the same, called a period, this time-period making it possible to determine without ambiguity the element under consideration. These periods, which can be measured by diverse methods, vary greatly. The period of uranium is several billions of years; that of radium is about 1600 years; that of its emanation a little less than four days; and there are among the following descendants some whose period is the small fraction of a second. The exponential law has a profound philosophic bearing; it indicates that the transformation is produced according to the laws of probability. The causes that determine the transformation are a mystery to us, and we do not yet know if they derive from causal conditions outside the atom, or from conditions of internal instability. In many cases, up to the present, no exterior action has shown itself effective in influencing the transformation.

This rapid succession of discoveries which overthrew familiar scientific conceptions long held in physics and chemistry did not fail to meet, at first, with doubts and incredulity. But the great part of the scientific world received them with enthusiasm. At the same time Pierre Curie's fame grew in France and in foreign countries. Already in 1901 the Academy of Sciences had awarded him the Lacaze prize. In 1902, Mascart, who had many times given him most valuable aid, decided to propose him as a member of the Academy of Sciences. It was not easy for Pierre Curie to agree to this, believing, as he did, that the Academy should elect its members without the necessity of any preliminary solicitation or paying of calls. Nevertheless, because of the friendly insistence of Mascart, and above all because the Physics Section of the Academy had already declared itself unanimously in his favor, he presented himself. In spite of this, however, he failed of election, and it was only in 1905 that he became a member of the Institute, a membership which did not last even a year. He was also elected to several academies and scientific societies in other countries, and given an honorary doctor's degree by several universities.

During 1903 we went to London at the invitation of the Royal Institution, before which my husband was to lecture on radium. On this occasion he had a most enthusiastic reception. He was especially happy to see here again Lord Kelvin, who had always expressed an affection for him, and who, despite his advanced age, preserved an interest, perennially young, in science. The illustrious scientist showed, with touching satisfaction, a glass vial containing a grain of radium salt that Pierre Curie had given him. We met here also other celebrated scientists, as Crookes, Ramsay, and J. Dewar. In collaboration with the latter, Pierre Curie published investigations on the discharge of heat by radium at very low temperatures, and upon the formation of helium in radium salt.

A few months later the Davy medal was conferred upon him (and also upon me) by the Royal Society of London, and at almost the same time, we received, together with Henri Becquerel, the Nobel prize for physics. Our health prevented us from attending the ceremony for the awarding of this prize in December, and it was only in June, 1905, that we were able to go to Stockholm where Pierre Curie gave his Nobel lecture. We were most cordially received and had the felicity of seeing the admirable Swedish nature in its most brilliant aspect.

The award of the Nobel prize was an important event for us because of the prestige carried by the Nobel foundation, only recently founded (1901). Also, from a financial point of view, the half of the prize represented an important sum. It meant that in the future Pierre Curie could turn over his teaching in the School of Physics to Paul Langevin, one of his former students, and a physicist of great competence. He could also engage a preparator to aid him in his work.

But at the same time the publicity this very happy event entailed bore very heavily on a man who was neither prepared for it, nor accustomed to it. There followed an avalanche of visits, of letters, of demands for articles and lectures, which meant a constant enervation, fatigue, and loss of time. He was kind and did not like to refuse a request; but on the other hand, he had to recognize that he could not accede to the solicitations that overwhelmed him without disastrous results to his health, as well as to his peace of mind, and his work. In a letter to Ch. Ed. Guillaume, he said:

"People ask me for articles and lectures, and after a few years are passed, the very persons who make these demands will be astonished to see that we have not accomplished any work."

And in other letters of the same period, written to E. Gouy, he expressed himself as follows:

"20 March 1902

"As you have seen, fortune favors us at this moment; but these favors of fortune do not come without many worries. We have never been less tranquil than at this moment. There are days when we scarcely have time to breathe. And to think that we dreamed of living in the wild, quite removed from human beings!"

"22 January 1904

"MY DEAR FRIEND:

"I have wanted to write to you for a long time; excuse me if I have not done so. The cause is the stupid life which I lead at present. You have seen this sudden infatuation for radium, which has resulted for us in all the advantages of a moment of popularity. We have been pursued by journalists and photographers from all countries of the world; they have gone even so far as to report the conversation between my daughter and her nurse, and to describe the black-and-white cat that lives with us.... Further, we have had a great many appeals for money.... Finally, the collectors of autographs, snobs, society people, and even at times, scientists, have come to see us—in our magnificent and tranquil quarters in the laboratory—and every evening there has been a voluminous correspondence to send off. With such a state of things I feel myself invaded by a kind of stupor. And yet all this turmoil will not perhaps have been in vain, if it results in my getting a chair and a laboratory. To tell the truth, it will be necessary to create the chair, and I shall not have the laboratory at first. I should have preferred the reverse, but Liard wishes to take advantage of the present moment to bring about the creation of a new chair that will later be acquired for the university. They are to establish a chair without a fixed program, which will be something like a course in the Collège de France, and I believe I shall be obliged to change my subject each year, which will be a great trial to me."

"31 January 1905

"... I have had to give up going to Sweden. We are, as you see, most irregular in our relations with the Swedish Academy; but, to tell the truth, I can only keep up by avoiding all physical fatigue. And my wife is in the same condition; we can no longer dream of the great work days of times gone by.

"As to research, I am doing nothing at present. With my course, my students, apparatus to install, and the interminable procession of people who come to disturb me without serious reason, the days pass without my having been able to achieve anything useful at this end."

"25 July 1905

"MY DEAR FRIEND:

"We have regretted so much being deprived of your visit this year, but hope to see you in October. If we do not make an effort from time to time, we end by losing touch with our best and most congenial friends, and in keeping company with others for the simple reason that it is easy to meet them.

"We continue to lead the same life of people who are extremely occupied, without being able to accomplish anything interesting. It is now more than a year since I have been able to engage in any research, and I have no moment to myself. Clearly I have not yet discovered a means to defend ourselves against this frittering away of our time which is nevertheless extremely necessary. Intellectually, it is a question of life or death."

"7 November 1905

"I begin my course tomorrow but under very bad conditions for the preparation of my experiments. The lecture room is at the Sorbonne, and my laboratory is in the rue Cuvier. Besides, a great number of other courses are given in the same lecture room, and I can use it only one morning for the preparation of my own.

"I am neither very well, nor very ill; but I am easily fatigued, and I have left but very little capacity for work. My wife, on the contrary, leads a very active life, between her children, the School at Sèvres, and the laboratory. She does not lose a minute, and occupies herself more regularly than I can with the direction of the laboratory in which she passes the greater part of the day."

To sum up: despite these outside complications, our life, by a common effort of will, remained as simple and as retired as formerly. Toward the close of 1904 our family was increased by the birth of a second daughter. Eve Denise was born in the modest house in Boulevard Kellermann, where we still lived with Doctor Curie, seeing only a few friends.

As our elder daughter grew up, she began to be a little companion to her father, who took a lively interest in her education and gladly went for walks with her in his free times, especially on his vacation days. He carried on serious conversations with her, replying to all her questions and delighting in the progressive development of her young mind. From their early age, his children enjoyed his tender affection, and he never wearied of trying to understand these little beings, in order to be able to give them the best he had to give.

With his great success in other countries, the complete appreciation of Pierre Curie in France, however tardily, did at last follow. At forty-five he found himself in the first rank of French scientists and yet, as a teacher, he occupied an inferior position. This abnormal state of affairs aroused public opinion in his favor, and under the influence of this wave of feeling, the director of the Academy of Paris, L. Liard, asked Parliament to create a new professorship in the Sorbonne, and at the beginning of the academic year 1904-05 Pierre Curie was named titular professor of the Faculty of Sciences of Paris. A year later he definitely quitted the School of Physics where his substitute, Paul Langevin, succeeded him.

This new professorship was not established without a few difficulties. The first project had provided for a new chair, but not for a laboratory. And Pierre Curie felt that he could not accept a situation which involved the risk of losing even the mediocre means of work that he then had, instead of offering better ones. He wrote, therefore, to his chiefs, that he had decided to remain at the P.C.N. His firmness won the day. To the new chair was added a fund for a laboratory and personnel for the new work (a chief of laboratory, a preparator, and a laboratory boy). The position of chief of laboratory was offered to me, which was a cause of very great satisfaction to my husband.

It was not without regret that we left the School of Physics, where we had known such happy work days, despite their attendant difficulties. We had become particularly attached to our hangar, which continued to stand, though in a state of increasing decay, for several years, and we went to visit it from time to time. Later it had to be pulled down to make way for a new building for the Physics School, but we have preserved photographs of it. Warned of its approaching destruction by the faithful Petit, I made my last pilgrimage there, alas, alone. On the blackboard there was still the writing of him who had been the soul of the place; the humble refuge for his research was all impregnated with his memory. The cruel reality seemed some bad dream; I almost expected to see the tall figure appear, and to hear the sound of the familiar voice.

Even though Parliament had voted the creation of a new chair, it did not go so far as to consider the simultaneous founding of a laboratory which was, nevertheless, necessary to the development of the new science of radioactivity. Pierre Curie therefore kept the little workroom at the P.C.N., and secured as a temporary solution of his difficulty the use of a large room, then not being used by the P.C.N. He arranged, too, to have a little building consisting of two rooms and a study set up in the court.

One cannot help feeling sorrow in realizing that this was a last concession, and that actually one of the first French scientists never had an adequate laboratory to work in, and this even though his genius had revealed itself as early as his twentieth year. Without doubt if he had lived longer, he would have had the benefit of satisfactory conditions for his work, but he was still deprived of them at his death at the premature age of forty-eight. Can we fully imagine the regret of an enthusiastic and disinterested worker in a great work, who is retarded in the realization of his dream by the constant lack of means? And can we think without a feeling of profound grief of the waste—

the one irreparable one—of the nation's greatest asset: the genius, the powers, and the courage of its best children?

Pierre Curie had always in mind his urgent need for a good laboratory. When, because of his great reputation, his chiefs felt obliged to try to induce him, in 1903, to accept the decoration of the Légion d'Honneur, he declined that distinction, remaining true to the opinion already referred to in a preceding chapter. And the letter he wrote on this occasion was inspired by the same feeling as that in the one previously quoted, when he wrote to his director to refuse the palmes académiques. I quote an extract:

"I pray you to thank the Minister, and to inform him that I do not in the least feel the need of a decoration, but that I do feel the greatest need for a laboratory."

After he was named professor at the Sorbonne, Pierre Curie had to prepare a new course. The position had been given a very personal character and a very general scope. He was left great freedom in the choice of the matter he would present. Taking advantage of this freedom he returned to a subject that was dear to him, and devoted part of his lectures to the laws of symmetry, the study of fields of vectors and tensors, and to the application of these ideas to the physics of crystals. He intended to carry these lessons further, and to work out a course that would completely cover the physics of crystallized matter which would have been especially useful because this subject was so little known in France. His other lessons dealt with radioactivity, set forth the discoveries made in this new domain, and the revolution they had caused in science.

Even though he was very much absorbed in the preparation of his course, and often ill, my husband continued, nevertheless, to work in the laboratory, which was becoming better and better organized. He had a little more space now, and could receive a few students. In collaboration with A. Laborde, he carried on investigations in mineral waters and gases discharged from springs. This was the last work he published.

His intellectual faculties were at this time at their height. One could but admire the surety and rigor of his reasoning on the theories of physics, his clear comprehension of fundamental principles, and a certain profound sense of phenomena which he had by instinct, but which he perfected during the course of a life entirely consecrated to research and reflection. His skill in experiment, remarkable from the beginning, was increased by practice. He experienced the pleasure of an artist when he succeeded with a delicate installation. He enjoyed, too, devising and constructing new apparatus, and I used jokingly to tell him that he would not be happy unless he made at least an attempt of this kind once every six months. His natural curiosity and vivid imagination pushed him to undertakings in very varied directions; he could change the object of his research with surprising ease.

He was scrupulously careful of scientific probity and of complete accuracy in his publications. These are very perfect in form, and none the less so in those parts where he applies the critical spirit to himself, expressing his determination never to affirm anything that does not seem entirely clear. He expresses his thought on this point in the following words:

"In the study of unknown phenomena, one can make very general hypotheses and then advance step by step with the help of experience. This method of progress is sure but necessarily slow. One can, on the contrary, make daring hypotheses in which he specifies the mechanism of phenomena. Such a method of procedure has the advantage of suggesting certain experiments, and, above all, of facilitating reasoning by rendering it less abstract through the employment of an image. But on the other hand, one cannot hope thus to conceive a complex theory in accord with experiment. The precise hypothesis almost certainly includes a portion of error along with a portion of truth. And this last portion, if it exists, forms only a part of a more general proposition to which it will be necessary in the end to return."

Moreover, even though he never hesitated to make hypotheses, he never permitted their premature publication. He could never accustom himself to a system of work which involved hasty publications, and was always happier in a domain in which but a few investigators were quietly working. The considerable vogue of radioactivity made him wish to abandon this field of research for a time, and to return to his interrupted studies of the physics of crystals. He dreamed also of making an examination of diverse theoretical questions.

He gave much thought to his teaching, which constantly improved, and which suggested to him ideas on the general orientation of studies and on methods of teaching, which he believed should be based on contact with experience and nature. He hoped to see his views adopted by the Association of Professors as soon as it was formed, and to obtain the declaration "that the teaching of the sciences must be the dominant teaching of both the boys' and girls' lycées."

"But," he said, "such a notion would have little chance of success."

But this last period of his life, so fecund, was, alas, soon to end. His admirable scientific career was to be suddenly broken at the very moment when he could hope that the years of work to come would be less hard than those which had preceded.

In 1906, quite ill and tired, he went with me and the children to spend Easter in the Chevreuse Valley. Those were two sweet days under a mild sun, and Pierre Curie felt the weight of weariness lighten in a healing repose near to those who were dear to him. He amused himself in the meadows with his little girls, and talked with me of their present and their future.

He returned to Paris for a reunion and dinner of the Physics Society. There he sat beside Henri Poincaré and had a long conversation with him on methods of teaching. As we were returning on foot to our house, he continued to develop his ideas on the culture that he dreamed of, happy in the consciousness that I shared his views.

The following day, the 19th of April, 1906, he attended a reunion of the Association of Professors of the Faculties of the Sciences, where he talked with them very cordially about the aims which the Association might adopt. As he went out from this reunion and was crossing the rue Dauphine, he was struck by a truck coming from the

Pont Neuf, and fell under its wheels. A concussion of the brain brought instantaneous death.

So perished the hope founded on the wonderful being who thus ceased to he. In the study room to which he was never to return, the water buttercups he had brought from the country were still fresh.

[9]During my recent visit to America, where a gramme of radium was generously offered me by American women, the Buffalo Society of Natural Sciences presented me, as a souvenir, with a publication reviewing the development of the radium industry in the United States. This included photographic reproductions of letters from Pierre Curie in which he replied in as complete a manner as possible to the questions asked by American engineers. (1902 and 1903.)

[10]The price of a milligramme of the element of radium was then fixed at about 750 francs.

[11]These physicians were aided by the manufacturer, Armet de Lisle, who placed at their disposition the radium needed for their first undertakings. He founded, besides, in 1906, a laboratory for clinical study, provided with a supply of radium. And he subventioned the first special publication devoted to radioactivity and its applications, as a journal under the name Radium, edited by J. Danne. This is an example of generous support of science by industry, in reality still very rare but which one wishes might become general, in the common interest of these two branches of human activity.

[12]The hypothesis according to which radioactivity is bound up with the atomic transformation of elements was first envisaged by Pierre Curie and by me, along with other possible hypotheses, before it was utilized by E. Rutherford. (See Revue Scientifique, 1900, Mme. Curie, etc.)

[13]By using the unusual energy of Alpha-rays E. Rutherford has obtained recently the rupture of certain light atoms, like those of nitrogen.

# CHAPTER VII

## THE NATION'S SORROW. THE LABORATORIES: "SACRED PLACES"

I shall not attempt to describe the grief of the family left by Pierre Curie. By what I have earlier said in this narrative one can understand what he meant to his father, his brother, and his wife. He was, too, a devoted father, tender in his love for his children, and happy to occupy himself with them. But our daughters were still too young at this time to realize the calamity that had befallen us. Their grandfather and I, united in our common suffering, did what we could to see that their childhood should not be too much darkened by the disaster.

The news of the catastrophe caused veritable consternation in the scientific world of France, as well as in that of other countries. The heads of the university and the professors expressed their emotion in letters full of sympathy, and a great number of foreign scientists also sent letters and telegrams. No less deep was the impression produced on the public with whom Pierre Curie, despite his reserve, enjoyed great renown. This feeling was expressed in numerous private letters coming not only from those whom we knew, but also from persons entirely unknown to us. At the same time the press printed articles of regret, bearing the stamp of deep sincerity. The French government sent its condolences, and a few rulers of foreign countries sent their personal expressions of sympathy. One of the purest glories of France had been extinguished, and each understood that this was a nation's sorrow.[14]

Faithful to the memory of him who had left us, we wished a simple interment in the family vault in the little cemetery at Sceaux. There was neither official ceremony nor address, and only his friends accompanied him to his last home. As he thought of him who was no more, his brother Jacques said to me: "He had all the gifts; there were not two like him."

In order to assure the continuance of his work, the Faculty of Sciences of Paris paid me the very great honor of asking me to take the place that he had occupied. I accepted this heavy heritage, in the hope that I might build up some day, in his memory, a laboratory worthy of him, which he had never had, but where others would be able to work to develop his idea. This hope is now partly realized, thanks to the common initiative of the University and the Pasteur Institute, which have aimed at the creation of a Radium Institute, composed of two laboratories, the Curie and the Pasteur, destined for the physicochemical and the biological study of radium rays. In touching homage to him who had disappeared the new street leading to the Institute was named rue Pierre Curie.

This Institute is, however, insufficient in view of the considerable development of radioactivity and of its therapeutic applications. The best authorized persons now recognize that France must possess a Radium Institute comparable to those of England and America for the Curietherapie which has become an efficacious means in the battle against cancer. It is to be hoped that with generous and far-seeing aid, we shall have, in a few years, a Radium Institute complete and enlarged, worthy of our country.

To honor the memory of Pierre Curie, the French Society of Physics decided to issue a complete publication of his works. This publication, arranged by P. Langevin and C. Cheneveau, comprises but a single volume of about 600 pages, which appeared in 1908, and for which I wrote a preface. This unique volume, which includes a work as important as it is varied, is a faithful reflection of the mentality of the author. One finds in it a great richness of ideas and of experimental facts leading to clear and well-established results, but the exposition is limited to the strictly necessary, and is irreproachable, one might even say classical, in form. It is to be regretted that Pierre Curie did not use his gifts as scientist and author in writing extended memoirs or books. It was not the desire that was lacking; he had several cherished projects of this nature. But he could never put them into execution because of the difficulties with which he had to struggle during all his working life.

And now, let us glance at this narrative as a whole, in which I have attempted to evoke the image of a man who, inflexibly devoted to the service of his ideal, honored humanity by an existence lived in silence, in the simple grandeur of his genius and his character. He had the faith of those who open new ways. He knew that he had a high mission to fulfil and the mystic dream of his youth pushed him invincibly beyond the usual path of life into a way which he called anti-natural because it signified the renunciation of the pleasures of life. Nevertheless, he resolutely subordinated his thoughts and desires to this dream, adapting himself to it and identifying himself with it more and more completely. Believing only in the pacific might of science and of reason, he lived for the search of truth. Without prejudice or parti pris, he carried the same loyalty into his study of things that he used in his understanding of other men and of himself. Detached from every common passion, seeking neither supremacy nor honors, he had no enemies, even though the effort he had achieved in the control of himself had made of him one of those elect whom we find in advance of their time in all the epochs of civilization. Like them he was able to exercise a profound influence merely by the radiation of his inner strength.

It is useful to learn how much sacrifice such a life represents. The life of a great scientist in his laboratory is not, as many may think, a peaceful idyll. More often it is a bitter battle with things, with one's surroundings, and above all with oneself. A great discovery does not leap completely achieved from the brain of the scientist, as Minerva sprang, all panoplied, from the head of Jupiter; it is the fruit of accumulated preliminary work. Between the days of fecund productivity are inserted days of uncertainty when nothing seems to succeed, and when even matter itself seems hostile, and it is then that one must hold out against discouragement. Thus without ever forsaking his inexhaustible patience, Pierre Curie used sometimes to say to me: "It is nevertheless hard, this life that we have chosen."

For the admirable gift of himself, and for the magnificent service he renders humanity, what reward does our society offer the scientist? Have these servants of an idea the necessary means of work? Have they an assured existence, sheltered from care? The example of Pierre Curie, and of others, shows that they have none of these things; and that more often, before they can secure possible working conditions, they have to exhaust their youth and their powers in daily anxieties. Our society, in which reigns an eager desire for riches and luxury, does not understand the value of science. It does not realize that science is a most precious part of its moral patrimony. Nor does it take sufficient cognizance of the fact that science is at the base of all the progress that lightens the burden of life and lessens its suffering. Neither public powers nor private generosity actually accord to science and to scientists the support and the subsidies indispensable to fully effective work.

I invoke, in closing, the admirable pleading of Pasteur:

"If the conquests useful for humanity touch your heart, if you are overwhelmed before the astonishing results of electric telegraphy, of the daguerrotype, of anesthesia, and of other wonderful discoveries, if you are jealous of the part your country may claim in the spreading of these marvelous things, take an interest, I beg of you, in those sacred places to which we give the expressive name of laboratories. Demand that they be multiplied and ornamented, for these are the temples of the future, of wealth, and of well-being. It is in them that humanity grows, fortifies itself, and becomes better. There it may learn to read in the works of nature the story of progress and of universal harmony, even while its own creations are too often those of barbarism, fanaticism, and destruction."

May this truth be widely spread, and deeply penetrate public opinion, that the future may be less hard for the pioneers who must open up new domains for the general good of humanity.

Extracts from Published Appreciations

I have chosen certain extracts from various published appreciations of Pierre Curie in order to complete my account by a few moving testimonies from eminent men of science.

Henri Poincaré:

"Curie was one of those on whom Science and France believed they had the right to count. His age permitted far-reaching hopes; what he had already given seemed a promise, and we knew that, living, he would not have failed. On the night preceding his death (pardon this personal memory) I sat next to him and he talked with me of his plans and his ideas. I admired the fecundity and the depth of his thought, the new aspect which physical phenomena took on when looked at through that original and lucid mind. I felt that I better understood the grandeur of human intelligence—and the following day, in an instant, all was annihilated. A stupid accident brutally reminded us how little place thought holds in the face of the thousand blind forces that hurl themselves across the world without knowing whither they go, crushing all in their passage.

"His friends, his colleagues understood at once the import of the loss they suffered, but the grief extended far beyond them. In foreign countries the most illustrious scientists joined in trying to show the esteem in which they held our compatriot, while in our own land there was no Frenchman, however ignorant, who did not feel more or less vaguely what a force his nation and humanity had lost.

"Curie brought to his study of physical phenomena I do not know what very fine sense which made him divine unsuspected analogies, and made it possible for him to orient himself in a labyrinth of complex appearances where others would have gone astray.... True physicists, like Curie, neither look within themselves, nor on the surface of things, but they know how to look through things.

"All those who knew him knew their pleasure and surety in his acquaintance, and the delicate charm that was exhaled, one might say, by his gentle modesty, by his naïve directness, by the fineness of his spirit. Always ready to efface himself before his family, before his friends, and even before his rivals, he was what one calls a 'poor candidate'; but in our democracy candidates are the least thing we lack.

"Who would have thought that so much gentleness concealed an intransigeant soul? He did not compromise with those general principles on which he was nourished, nor with the particular moral ideal he had been taught to love, that ideal of absolute sincerity, too high, perhaps, for the world in which we live. He did not know the thousand little accommodations with which our weakness contents itself. Moreover, he never separated the worship of this ideal from what he rendered to science, and he gave us a shining example of the high conception of duty that may spring from a simple and pure love of truth. It matters little in what God he believed; it is not the God, but faith, that performs miracles."

Institut de France: Written about P. Curie by M. D. Gernez.

"All for work, all for science: this sums up the life of Pierre Curie, a life so rich in brilliant discoveries and in the outlook of genius that it won him practically universal admiration. In the full maturity of his investigations whose progress he so eagerly pursued his work was ended, to the consternation of us all, by a terrible catastrophe on the 19th of April, 1906....

"All these honors did not dazzle him; he was and he will remain a remarkable figure among those who make the scientific history of our epoch. His contemporaries found in him a precious example of a devotion to science at once unyielding and disinterested. There have been few lives more pure and more justly famous."

Jean Perrin:

"Pierre Curie, whom all called a master, and whom we had the joy to call, too, our friend, died suddenly in the fullness of his powers.... We will try to show through him, as an example, what part a powerful genius can return to sincerity, to liberty, to the strong and calm audacity of thought which nothing can enchain and nothing can astonish. We acknowledge also all the greatness of the soul where these fine qualities of intelligence and character were united in a most noble unselfishness and most exquisite goodness.

"Those who have known Pierre Curie, know that, near him one felt awaken the need to do and to understand. We will try to honor his memory by spreading abroad this impression, and we will ask his pale and beautiful face for the secret of that radiation which made all those who approached him better men."

C. Cheneveau:

"... In order to realize our irreparable loss we must remember Curie's attachment to his students.... Some of us offered him, with reason, a veritable worship.... For myself, he was, next to my own family, one of those I loved most. How well he knew how to surround his simple collaborator with a great and tender affection. His immense kindness extended even to his most humble helpers, who adored him. I have never seen more sincere and more heart-breaking tears than those shed by the laboratory boys on the news of his sudden death."

Paul Langevin:

"... The hours when one could meet him and in which one loved to talk about his science and in which one thought with him, return each day to recall his memory, to bring back his kindly and thoughtful face, his luminous eyes and his beautiful, expressive head modeled by twenty-five years passed in the laboratory, and by a life of unremittent work and complete simplicity.

"... It is in his laboratory that my memories, still so recent, most readily bring him back to me, as he would appear to those near to whom he had grown older, scarcely changed by the eighteen years that have passed since. Timid and often awkward, I began under him my laboratory education....

"Surrounded by apparatus for the greater part conceived or modified by himself, he manipulated it with extreme dexterity, with the familiar gestures of the long white hands of the physicist....

"He was twenty-nine years old when I entered as a student. The mastery which ten years, passed entirely in the laboratory, had given him, imposed itself even on us, despite our ignorance, by the surety of his movements and explanations, and the ease, shaded by timidity, of his manner. We returned always with joy to the laboratory, where it was good to work near him because we felt him working near to us in that large, light room filled with apparatus whose forms were still a little mysterious to us. We did not fear to enter it often to consult him, and he sometimes admitted us, too, to perform some particularly delicate manipulation. Probably my finest memories of my school years are those of moments passed there standing before the blackboard where he took pleasure in talking with us, in awakening in us fruitful ideas, and in discussions of research which formed our taste for the things of science. His live and contagious curiosity, the fullness and surety of his information made him an admirable awakener of spirits."

I have wished above all, in gathering together here these few memories, in a bouquet reverently placed upon his tomb, to help, if I can, to fix the image of a man truly great in character and in thought, of a wonderful representative of the genius of our race. Entirely unfranchised from ancient servitudes, and passionately loving reason and clarity, he was an example—as is a prophet inspired by truths of the future—of what may be realized in moral beauty and goodness by a free and upright spirit, of constant courage, and of a mental honesty which made him repulse what he did not understand, and place his life in accord with this dream.

[14]From the great number of letters and telegrams of condolence, I quote, as examples, these lines written by three great scientists, today no longer living.

From M. Berthelot:

"MADAME:

"I do not wish to wait longer without sending you the sympathetic expression of my profound grief and of that of French and foreign scientists on the occasion of the common loss with you that we have all experienced. We were struck as by lightning by the tragic news! So many services already rendered science and humanity, so many services that we awaited from that genial inventor: all this vanished in an instant, or become already but a memory!"

From G. Lippmann:

"MADAME:

It is while traveling, and very late, that I receive the terrible news. I feel as if I had lost a brother; I did not know by what close ties I was attached to your husband. I know today. I suffer also for you, Madame. Believe in my sincere and respectful devotion."

From Lord Kelvin:

"Grievously distressed by terrible news of Curie's death. When will be funeral. We arrive Hotel Mirabeau tomorrow morning. Kelvin, Villa St. Martin, Cannes."

# AUTOBIOGRAPHICAL NOTES—MARIE CURIE

## CHAPTER I

I have been asked by my American friends to write the story of my life. At first, the idea seemed alien to me, but I yielded to persuasion. However, I could not conceive my biography as a complete expression of personal feelings or a detailed description of all incidents I would remember. Many of our feelings change with the years, and, when faded away, may seem altogether strange; incidents lose their momentary interest and may be remembered as if they have occurred to some other person. But there may be in a life some general direction, some continuous thread, due to a few dominant ideas and a few strong feelings, that explain the life and are characteristic of a human personality. Of my life, which has not been easy on the whole, I have described the general course and the essential features, and I trust that my story gives an understanding of the state of mind in which I have lived and worked.

My family is of Polish origin, and my name is Marie Sklodowska. My father and my mother both came from among the small Polish landed proprietors. In my country this class is composed of a large number of families, owners of small and medium-sized estates, frequently interrelated. It has been, until recently, chiefly from this group that Poland has drawn her intellectual recruits.

While my paternal grandfather had divided his time between agriculture and directing a provincial college, my father, more strongly drawn to study, followed the course of the University of Petrograd, and later definitely established himself at Warsaw as Professor of Physics and Mathematics in one of the lyceums of that city. He married a young woman whose mode of life was congenial to his; for, although very young, she had, what was, for that time, a very serious education, and was the director of one of the best Warsaw schools for young girls.

My father and mother worshiped their profession in the highest degree and have left, all over their country, a lasting remembrance with their pupils. I cannot, even to-day, go into Polish society without meeting persons who have tender memories of my parents.

Although my parents adopted a university career, they continued to keep in close touch with their numerous family in the country. It was with their relatives that I frequently spent my vacation, living in all freedom and finding opportunities to know the field life by which I was deeply attracted. To these conditions, so different from the usual villegiature, I believe, I owe my love for the country and nature.

Born at Warsaw, on the 7th of November, 1867, I was the last of five children, but my oldest sister died at the early age of fourteen, and we were left, three sisters and a brother. Cruelly struck by the loss of her daughter and worn away by a grave illness, my mother died at forty-two, leaving her husband in the deepest sorrow with his children. I was then only nine years old, and my eldest brother was hardly thirteen.

This catastrophe was the first great sorrow of my life and threw me into a profound depression. My mother had an exceptional personality. With all her intellectuality she had a big heart and a very high sense of duty. And, though possessing infinite indulgence and good nature, she still held in the family a remarkable moral authority. She had an ardent piety (my parents were both Catholics), but she was never intolerant; differences in religious belief did not trouble her; she was equally kind to any one not sharing her opinions. Her influence over me was extraordinary, for in me the natural love of the little girl for her mother was united with a passionate admiration.

Very much affected by the death of my mother, my father devoted himself entirely to his work and to the care of our education. His professional obligations were heavy and left him little leisure time. For many years we all felt weighing on us the loss of the one who had been the soul of the house.

We all started our studies very young. I was only six years old, and, because I was the youngest and smallest in the class, was frequently brought forward to recite when there were visitors. This was a great trial to me, because of my timidity; I wanted always to run away and hide. My father, an excellent educator, was interested in our work and knew how to direct it, but the conditions of our education were difficult. We began our studies in private schools and finished them in those of the government.

Warsaw was then under Russian domination, and one of the worst aspects of this control was the oppression exerted on the school and the child. The private schools directed by Poles were closely watched by the police and overburdened with the necessity of teaching the Russian language even to children so young that they could scarcely speak their native Polish. Nevertheless, since the teachers were nearly all of Polish nationality, they endeavored in every possible way to mitigate the difficulties resulting from the national persecution. These schools, however, could not legally give diplomas, which were obtainable only in those of the government.

The latter, entirely Russian, were directly opposed to the Polish national spirit. All instruction was given in Russian, by Russian professors, who, being hostile to the Polish nation, treated their pupils as enemies. Men of moral and intellectual distinction could scarcely agree to teach in schools where an alien attitude was forced upon them. So what the pupils were taught was of questionable value, and the moral atmosphere was altogether unbearable. Constantly held in suspicion and spied upon, the children knew that a single conversation in Polish, or an imprudent word, might seriously harm, not only themselves, but also their families. Amidst these hostilities, they lost all the joy of life, and precocious feelings of distrust and indignation weighed upon their childhood. On the other side, this abnormal situation resulted in exciting the patriotic feeling of Polish youths to the highest degree.

Yet of this period of my early youth, darkened though it was by mourning and the sorrow of oppression, I still keep more than one pleasant remembrance. In our quiet but occupied life, reunions of relatives and friends of our family brought some joy. My father was very interested in literature and well acquainted with Polish and foreign poetry; he even composed poetry himself and was able to translate it from foreign languages into Polish in a very successful way. His little poems on family events were our delight. On Saturday evenings he used to recite or read to us the masterpieces of Polish prose and poetry. These evenings were for us a great pleasure and a source of renewed patriotic feelings.

Since my childhood I have had a strong taste for poetry, and I willingly learned by heart long passages from our great poets, the favorite ones being Mickiewecz, Krasinski and Slowacki. This taste was even more developed when I became acquainted with foreign literatures; my early studies included the knowledge of French, German, and Russian, and I soon became familiar with the fine works written in these languages. Later I felt the need of knowing English and succeeded in acquiring the knowledge of that language and its literature.

My musical studies have been very scarce. My mother was a musician and had a beautiful voice. She wanted us to have musical training. After her death, having no more encouragement from her, I soon abandoned this effort, which I often regretted afterwards.

I learned easily mathematics and physics, as far as these sciences were taken in consideration in the school. I found in this ready help from my father, who loved science and had to teach it himself. He enjoyed any explanation he could give us about Nature and her ways. Unhappily, he had no laboratory and could not perform experiments.

The periods of vacations were particularly comforting, when, escaping the strict watch of the police in the city, we took refuge with relatives or friends in the country. There we found the free life of the old-fashioned family estate; races in the woods and joyous participation in work in the far-stretching, level grain-fields. At other times we passed the border of our Russian-ruled division (Congress Poland) and went southwards into the mountain country of Galicia, where the Austrian political control was less oppressive than that which we suffered. There we could speak Polish in all freedom and sing patriotic songs without going to prison.

My first impression of the mountains was very vivid, because I had been brought up in the plains. So I enjoyed immensely our life in the Carpathian villages, the view of the pikes, the excursions to the valleys and to the high mountain lakes with picturesque names such as: "The Eye of the Sea." However, I never lost my attachment to the open horizon and the gentle views of a plain hill country.

Later I had the opportunity to spend a vacation with my father far more south in Podolia, and to have the first view of the sea at Odessa, and afterwards at the Baltic shore. This was a thrilling experience. But it was in France that I become acquainted with the big waves of the ocean and the ever-changing tide. All my life through, the new sights of Nature made me rejoice like a child.

Thus passed the period of our school life. We all had much facility for intellectual work. My brother, Doctor Sklodowski, having finished his medical studies, became later the chief physician in one of the principal Warsaw hospitals. My sisters and I intended to take up teaching as our parents had done. However, my elder sister, when grown up, changed her mind and decided to study medicine. She took the degree of doctor at the Paris University, married Doctor Dluski, a Polish physician, and together they established an important sanatorium in a wonderfully beautiful Carpathian mountain place of Austrian Poland. My second sister, married in Warsaw, Mrs. Szalay, was for many years a teacher in the schools, where she rendered great service. Later she was appointed in one of the lyceums of free Poland.

I was but fifteen when I finished my high-school studies, always having held first rank in my class. The fatigue of growth and study compelled me to take almost a year's rest in the country. I then returned to my father in Warsaw, hoping to teach in the free schools. But family circumstances obliged me to change my decision. My father, now aged and tired, needed rest; his fortune was very modest. So I resolved to accept a position as governess for several children. Thus, when scarcely seventeen, I left my father's house to begin an independent life.

That going away remains one of the most vivid memories of my youth. My heart was heavy as I climbed into the railway car. It was to carry me for several hours, away from those I loved. And after the railway journey I must drive for five hours longer. What experience was awaiting me? So I questioned as I sat close to the car window looking out across the wide plains.

The father of the family to which I went was an agriculturist. His oldest daughter was about my age, and although working with me, was my companion rather than my pupil. There were two younger children, a boy and a girl. My relations with my pupils were friendly; after our lessons we went together for daily walks. Loving the country, I did not feel lonesome, and although this particular country was not especially picturesque, I was satisfied with it in all seasons. I took the greatest interest in the agricultural development of the estate where the methods were considered as models for the region. I knew the progressive details of the work, the distribution of crops in the fields; I eagerly followed the growth of the plants, and in the stables of the farm I knew the horses.

In winter the vast plains, covered with snow, were not lacking in charm, and we went for long sleigh rides. Sometimes we could hardly see the road. "Look out for the ditch!" I would call to the driver. "You are going straight into it," and "Never fear!" he would answer, as over we went! But these tumbles only added to the gayety of our excursions.

I remember the marvelous snow house we made one winter when the snow was very high in the fields; we could sit in it and look out across the rose-tinted snow plains. We also used to skate on the ice of the river and to watch the weather anxiously, to make sure that the ice was not going to give way, depriving us of our pleasure.

Since my duties with my pupils did not take up all my time, I organized a small class for the children of the village who could not be educated under the Russian government. In this the oldest daughter of the house aided me. We taught the little children and the girls who wished to come how to read and write, and we put in circulation Polish books which were appreciated, too, by the parents. Even this innocent work presented danger, as all initiative of this kind was forbidden by the government and might bring imprisonment or deportation to Siberia.

My evenings I generally devoted to study. I had heard that a few women had succeeded in following certain courses in Petrograd or in foreign countries, and I was determined to prepare myself by preliminary work to follow their example.

I had not yet decided what path I would choose. I was as much interested in literature and sociology as in science. However, during these years of isolated work, trying little by little to find my real preferences, I finally turned towards mathematics and physics, and resolutely undertook a serious preparation for future work. This work I proposed doing in Paris, and I hoped to save enough money to be able to live and work in that city for some time.

My solitary study was beset with difficulties. The scientific education I had received at the lyceum was very incomplete; it was well under the bachelorship program of a French lyceum; I tried to add to it in my own way, with the help of books picked up at random. This method could not be greatly productive, yet it was not without results. I acquired the habit of independent work, and learned a few things which were to be of use later on.

I had to modify my plans for the future when my eldest sister decided to go to Paris to study medicine. We had promised each other mutual aid, but our means did not permit of our leaving together. So I kept my position for three and a half years, and, having finished my work with my pupils, I returned to Warsaw, where a position, similar to the one I had left, was awaiting me.

I kept this new place for only a year and then went back to my father, who had retired some time before and was living alone. Together we passed an excellent year, he occupying himself with some literary work, while I increased our funds by giving private lessons. Meantime I continued my efforts to educate myself. This was no easy task under the Russian government of Warsaw; yet I found more opportunities than in the country. To my great joy, I was able, for the first time in my life, to find access to a laboratory: a small municipal physical laboratory directed by one of my cousins. I found little time to work there, except in the evenings and on Sundays, and was generally left to myself. I tried out various experiments described in treatises on physics and chemistry, and the results were sometimes unexpected. At times I would be encouraged by a little unhoped-for success, at others I would be in the deepest despair because of accidents and failures resulting from my inexperience. But on the whole, though I was taught that the way of progress is neither swift nor easy, this first trial confirmed in me the taste for experimental research in the fields of physics and chemistry.

Other means of instruction came to me through my being one of an enthusiastic group of young men and women of Warsaw, who united in a common desire to study, and whose activities were at the same time social and patriotic. It was one of those groups of Polish youths who believed that the hope of their country lay in a great effort to develop the intellectual and moral strength of the nation, and that such an effort would lead to a better national situation. The nearest purpose was to work at one's own instruction and to provide means of instruction for workmen and peasants. In accordance with this program we agreed among ourselves to give evening courses, each one teaching what he knew best. There is no need to say that this was a secret organization, which made everything extremely difficult. There were in our group very devoted young people who, as I still believe today, could do truly useful work.

I have a bright remembrance of the sympathetic intellectual and social companionship which I enjoyed at that time. Truly the means of action were poor and the results obtained could not be considerable; yet I still believe that the ideas which inspired us then are the only way to real social progress. You cannot hope to build a better world without improving the individuals. To that end each of us must work for his own improvement, and at the same time share a general responsibility for all humanity, our particular duty being to aid those to whom we think we can be most useful.

All the experiences of this period intensified my longing for further study. And, in his affection for me, my father, in spite of limited resources, helped me to hasten the execution of my early project. My sister had just married at Paris, and it was decided that I should go there to live with her. My father and I hoped that, once my studies were finished, we would again live happily together. Fate was to decide otherwise, since my marriage was to hold me in France. My father, who in his own youth had wished to do scientific work, was consoled in our separation by the progressive success of my work. I keep a tender memory of his kindness and disinterestedness. He lived with the family of my married brother, and, like an excellent grandfather, brought up the children. We had the sorrow of losing him in 1902, when he had just passed seventy.

So it was in November, 1891, at the age of twenty-four, that I was able to realize the dream that had been always present in my mind for several years.

When I arrived in Paris I was affectionately welcomed by my sister and brother-in-law, but I stayed with them only for a few months, for they lived in one of the outside quarters of Paris where my brother-in-law was beginning a medical practice, and I needed to get nearer to the schools. I was finally installed, like many other students of my country, in a modest little room for which I gathered some furniture. I kept to this way of living during the four years of my student life.

It would be impossible to tell of all the good these years brought to me. Undistracted by any outside occupation, I was entirely absorbed in the joy of learning and understanding. Yet, all the while, my living conditions were far from easy, my own funds being small and my family not having the means to aid me as they would have liked to do. However, my situation was not exceptional; it was the familiar experience of many of the Polish students whom I knew. The room I lived in was in a garret, very cold in winter, for it was insufficiently heated by a small stove which often lacked coal. During a particularly rigorous winter, it was not unusual for the water to freeze in the basin in the night; to be able to sleep I was obliged to pile all my clothes on the bedcovers. In the same room I prepared my meals with the aid of an alcohol lamp and a few kitchen utensils. These meals were often reduced to bread with a cup of chocolate, eggs or fruit. I had no help in housekeeping and I myself carried the little coal I used up the six flights.

This life, painful from certain points of view, had, for all that, a real charm for me. It gave me a very precious sense of liberty and independence. Unknown in Paris, I was lost in the great city, but the feeling of living there alone, taking care of myself without any aid, did not at all depress me. If sometimes I felt lonesome, my usual state of mind was one of calm and great moral satisfaction.

All my mind was centered on my studies, which, especially at the beginning, were difficult. In fact, I was insufficiently prepared to follow the physical science course at the Sorbonne, for, despite all my efforts, I had not succeeded in acquiring in Poland a preparation as complete as that of the French students following the same course. So I was obliged to supply this deficiency, especially in mathematics. I divided my time between courses, experimental work, and study in the library. In the evening I worked in my room, sometimes very late into the night. All that I saw and learned that was new delighted me. It was like a new world opened to me, the world of science, which I was at last permitted to know in all liberty.

I have pleasant memories of my relations with my student companions. Reserved and shy at the beginning, it was not long before I noticed that the students, nearly all of whom worked seriously, were disposed to be friendly. Our conversations about our studies deepened our interest in the problems we discussed.

Among the Polish students I did not have any companions in my studies. Nevertheless, my relations with their small colony had a certain intimacy. From time to time we would gather in one another's bare rooms, where we could talk over national questions and feel less isolated. We would also go for walks together, or attend public reunions, for we were all interested in politics. By the end of the first year, however, I was forced to give up these relationships, for I found that all my energy had to be concentrated on my studies, in order to achieve them as soon as possible. I was even obliged to devote most of my vacation time to mathematics.

My persistent efforts were not in vain. I was able to make up for the deficiency of my training and to pass examinations at the same time with the other students. I even had the satisfaction of graduating in first rank as "licenciée es sciences physiques" in 1893, and in second rank as "licenciée es sciences mathématiques" in 1894.

My brother-in-law, recalling later these years of work under the conditions I have just described, jokingly referred to them as "the heroic period of my sister-in-law's life." For myself, I shall always consider one of the best memories of my life that period of solitary years exclusively devoted to the studies, finally within my reach, for which I had waited so long.

It was in 1894 that I first met Pierre Curie. One of my compatriots, a professor at the University of Fribourg, having called upon me, invited me to his home, with a young physicist of Paris, whom he knew and esteemed highly. Upon entering the room I perceived, standing framed by the French window opening on the balcony, a tall young man with auburn hair and large, limpid eyes. I noticed the grave and gentle expression of his face, as well as a certain abandon in his attitude, suggesting the dreamer absorbed in his reflections. He showed me a simple cordiality and seemed to me very sympathetic. After that first interview he expressed the desire to see me again and to continue our conversation of that evening on scientific and social subjects in which he and I were both interested, and on which we seemed to have similar opinions.

Some time later, he came to me in my student room and we became good friends. He described to me his days, filled with work, and his dream of an existence entirely devoted to science. He was not long in asking me to share that existence, but I could not decide at once; I hesitated before a decision that meant abandoning my country and my family.

I went back to Poland for my vacation, without knowing whether or not I was to return to Paris. But circumstances permitted me again to take up my work there in the autumn of that year. I entered one of the physics laboratories at the Sorbonne, to begin experimental research in preparation for my doctor's thesis.

Again I saw Pierre Curie. Our work drew us closer and closer, until we were both convinced that neither of us could find a better life companion. So our marriage was decided upon and took place a little later, in July, 1895.

Pierre Curie had just received his doctor's degree and had been made professor in the School of Physics and Chemistry of the City of Paris. He was thirty-six years old, and already a physicist known and appreciated in France and abroad. Solely preoccupied with scientific investigation, he had paid little attention to his career, and his material resources were very modest. He lived at Sceaux, in the suburbs of Paris, with his old parents, whom he loved tenderly, and whom he described as "exquisite" the first time he spoke to me about them. In fact, they were so: the father was an elderly physician of high intellect and strong character, and the mother the most excellent of women, entirely devoted to her husband and her sons. Pierre's elder brother, who was then professor at the University of Montpellier, was always his best friend. So I had the privilege of entering into a family worthy of affection and esteem, and where I found the warmest welcome.

We were married in the simplest way. I wore no unusual dress on my marriage day, and only a few friends were present at the ceremony, but I had the joy of having my father and my second sister come from Poland.

We did not care for more than a quiet place in which to live and to work, and were happy to find a little apartment of three rooms with a beautiful view of a garden. A few pieces of furniture came to us from our parents. With a money gift from a relative we acquired two bicycles to take us out into the country.

# CHAPTER II

With my marriage there began for me a new existence entirely different from the solitary life that I had known during the preceding years. My husband and I were so closely united by our affection and our common work that we passed nearly all of our time together. I have only a few letters from him, for we were so little apart. My husband spent all the time he could spare from his teaching at his research work in the laboratory of the school in which he was professor and I obtained authorization to work with him.

Our living apartment was near the school, so we lost little time in going and coming. As our material resources were limited, I was obliged to attend to most of the housekeeping myself, particularly the preparation of meals. It was not easy to reconcile these household duties with my scientific work, yet, with good will, I managed it. The great thing was that we were alone together in the little home which gave us a peace and intimacy that were very enjoyable for us.

At the same time that I was working in the laboratory, I still had to take a few study courses, for I had decided to take part in the examination for a certificate that would allow me to teach young girls. If I succeeded in this, I would be entitled to be named professor. In August, 1896, after having devoted several months to preparation, I came out first in the examination.

Our principal distraction from the close work of the laboratory consisted in walks or bicycle rides in the country. My husband greatly enjoyed the out-of-doors and took great interest in the plants and animals of woods and meadows. Hardly a corner in the vicinity of Paris was unknown to him. I also loved the country and these excursions were a great joy for me as well as to him, relieving our mind from the tension of the scientific work. We used to bring home bunches of flowers. Sometimes we forgot all about the time and got back late at night. We visited regularly my husband's parents where our room was always ready.

In the vacation we went on longer outings by means of our bicycles. In this way we covered much ground in Auvergne and in the Cevennes and visited several regions at the seashore. We took a great delight in these long all-day excursions, arriving at night always in a new place. If we stayed in one place too long, my husband began to wish to get back to the laboratory. It is also in vacation time that we visited once my family in the Carpathian mountains. My husband learned some Polish in view of this journey to Poland.

But first of all in our life was our scientific work. My husband gave much care to the preparation of his courses, and I gave him some assistance in this, which, at the time, helped me in my education. However, most of our time was devoted to our laboratory researches.

My husband did not then have a private laboratory. He could, to some extent, use the laboratory of the school for his own work, but found more freedom by installing himself in some unused corner of the Physics School building. I thus learned from his example that one could work happily even in very insufficient quarters. At this time my husband was occupied with researches on crystals, while I undertook an investigation of the magnetic properties of steel. This work was completed and published in 1897.

In that same year the birth of our first daughter brought a great change in our life. A few weeks later my husband's mother died and his father came to live with us. We took a small house with a garden at the border of Paris and continued to occupy this house as long as my husband lived.

It became a serious problem how to take care of our little Irene and of our home without giving up my scientific work. Such a renunciation would have been very painful to me, and my husband would not even think of it; he used to say that he had got a wife made expressly for him to share all his preoccupations. Neither of us would contemplate abandoning what was so precious to both.

Of course we had to have a servant, but I personally saw to all the details of the child's care. While I was in the laboratory, she was in the care of her grandfather, who loved her tenderly and whose own life was made brighter by her. So the close union of our family enabled me to meet my obligations. Things were particularly difficult only in case of more exceptional events, such as a child's illness, when sleepless nights interrupted the normal course of life.

It can be easily understood that there was no place in our life for worldly relations. We saw but a few friends, scientific workers, like ourselves, with whom we talked in our home or in our garden, while I did some sewing for my little girl. We also maintained affectionate relations with my husband's brother and his family. But I was separated from all my relatives, as my sister had left Paris with her husband to live in Poland.

It was under this mode of quiet living, organized according to our desires, that we achieved the great work of our lives, work begun about the end of 1897 and lasting for many years.

I had decided on a theme for my doctorate. My attention had been drawn to the interesting experiments of Henri Becquerel on the salts of the rare metal uranium. Becquerel had shown that by placing some uranium salt on a photographic plate, covered with black paper, the plate would be affected as if light had fallen on it. The effect is produced by special rays which are emitted by the uranium salt and are different from ordinary luminous rays as they can pass through black paper. Becquerel also showed that these rays can discharge an electroscope. He at first thought that the uranium rays were produced as a result of exposing the uranium salt to light, but experiment showed that salts kept for several months in the dark continued the peculiar rays.

My husband and I were much excited by this new phenomenon, and I resolved to undertake the special study of it. It seemed to me that the first thing to do was to measure the phenomenon with precision. In this I decided to use that property of the rays which enabled them to discharge an electroscope. However, instead of the usual electroscope,

I used a more perfect apparatus. One of the models of the apparatus used by me for these first measurements is now in the College of Physicians and Surgeons in Philadelphia.

I was not long in obtaining interesting results. My determinations showed that the emission of the rays is an atomic property of the uranium, whatever the physical or chemical conditions of the salt were. Any substance containing uranium is as much more active in emitting rays, as it contains more of this element.

I then thought to find out if there were other substances possessing this remarkable property of uranium, and soon found that substances containing thorium behaved in a similar way, and that this behavior depended similarly on an atomic property of thorium. I was now about to undertake a detailed study of the uranium and thorium rays when I discovered a new interesting fact.

I had occasion to examine a certain number of minerals. A few of them showed activity; they were those containing either uranium or thorium. The activity of these minerals would have had nothing astonishing about it, if it had been in proportion to the quantities of uranium or thorium contained in them. But it was not so. Some of these minerals revealed an activity three or four times greater than that of uranium. I verified this surprising fact carefully, and could not doubt its truth. Speculating about the reason for this, there seemed to be but one explanation. There must be, I thought, some unknown substance, very active, in these minerals. My husband agreed with me and I urged that we search at once for this hypothetical substance, thinking that, with joined efforts, a result would be quickly obtained. Neither of us could foresee that in beginning this work we were to enter the path of a new science which we should follow for all our future.

Of course, I did not expect, even at the beginning, to find a new element in any large quantity, as the minerals had already been analyzed with some precision. At least, I thought there might be as much as one per cent of the unknown substance in the minerals. But the more we worked, the clearer we realized that the new radioactive element could exist only in quite minute proportion and that, in consequence, its activity must be very great. Would we have insisted, despite the scarcity of our means of research, if we had known the true proportion of what we were searching for, no one can tell; all that can be said now is that the constant progress of our work held us absorbed in a passionate research, while the difficulties were ever increasing. As a matter of fact, it was only after several years of most arduous labor that we finally succeeded in completely separating the new substance, now known to everybody as radium. Here is, briefly, the story of the search and discovery.

As we did not know, at the beginning, any of the chemical properties of the unknown substance, but only that it emits rays, it was by these rays that we had to search. We first undertook the analysis of a pitchblende from St. Joachimsthal. Analyzing this ore by the usual chemical methods, we added an examination of its different parts for radioactivity, by the use of our delicate electrical apparatus. This was the foundation of a new method of chemical analysis which, following our work, has been extended, with the result that a large number of radioactive elements have been discovered.

In a few weeks we could be convinced that our prevision had been right, for the activity was concentrating in a regular way. And, in a few months, we could separate from the pitchblende a substance accompanying the bismuth, much more active than uranium, and having well defined chemical properties. In July, 1898, we announced the existence of this new substance, to which I gave the name of polonium, in memory of my native country.

While engaged in this work on polonium, we had also discovered that, accompanying the barium separated from the pitchblende, there was another new element. After several months more of close work we were able to separate this second new substance, which was afterwards shown to be much more important than polonium. In December, 1898, we could announce the discovery of this new and now famous element, to which we gave the name of radium.

However, the greatest part of the material work had yet to be done. We had, to be sure, discovered the existence of the remarkable new elements, but it was chiefly by their radiant properties that these new substances were distinguished from the bismuth and barium with which they were mixed in minute quantities. We had still to separate them as pure elements. On this work we now started.

We were very poorly equipped with facilities for this purpose. It was necessary to subject large quantities of ore to careful chemical treatment. We had no money, no suitable laboratory, no personal help for our great and difficult undertaking. It was like creating something out of nothing, and if my earlier studying years had once been called by my brother-in-law the heroic period of my life, I can say without exaggeration that the period on which my husband and I now entered was truly the heroic one of our common life.

We knew by our experiments that in the treatment of pitchblende at the uranium plant of St. Joachimsthal, radium must have been left in the residues, and, with the permission of the Austrian government, which owned the plant, we succeeded in securing a certain quantity of these residues, then quite valueless,—and used them for extraction of radium. How glad I was when the sacks arrived, with the brown dust mixed with pine needles, and when the activity proved even greater than that of the primitive ore! It was a stroke of luck that the residues had not been thrown far away or disposed of in some way, but left in a heap in the pine wood near the plant. Some time later, the Austrian government, on the proposition of the Academy of Science of Vienna, let us have several tons of similar residues at a low price. With this material was prepared all the radium I had in my laboratory up to the date when I received the precious gift from the American women.

The School of Physics could give us no suitable premises, but for lack of anything better, the Director permitted us to use an abandoned shed which had been in service as a dissecting room of the School of Medicine. Its glass roof did not afford complete shelter against rain; the heat was suffocating in summer, and the bitter cold of winter was only a little lessened by the iron stove, except in its immediate vicinity. There was no question of obtaining the needed proper apparatus in common use by chemists. We simply had some old pine-wood tables with furnaces and gas burners. We had to use the adjoining yard for those of our chemical operations that involved producing irritating gases; even then the gas often filled our shed. With this equipment we entered on our exhausting work.

Yet it was in this miserable old shed that we passed the best and happiest years of our life, devoting our entire days to our work. Often I had to prepare our lunch in the shed, so as not to interrupt some particularly important operation. Sometimes I had to spend a whole day mixing a boiling mass with a heavy iron rod nearly as large as myself. I would be broken with fatigue at the day's end. Other days, on the contrary, the work would be a most minute and delicate fractional crystallization, in the effort to concentrate the radium. I was then annoyed by the floating dust of iron and coal from which I could not protect my precious products. But I shall never be able to express the joy of the untroubled quietness of this atmosphere of research and the excitement of actual progress with the confident hope of still better results. The feeling of discouragement that sometimes came after some unsuccessful toil did not last long and gave way to renewed activity. We had happy moments devoted to a quiet discussion of our work, walking around our shed.

One of our joys was to go into our workroom at night; we then perceived on all sides the feebly luminous silhouettes of the bottles or capsules containing our products. It was really a lovely sight and one always new to us. The glowing tubes looked like faint, fairy lights.

Thus the months passed, and our efforts, hardly interrupted by short vacations, brought forth more and more complete evidence. Our faith grew ever stronger, and our work being more and more known, we found means to get new quantities of raw material and to carry on some of our crude processes in a factory, allowing me to give more time to the delicate finishing treatment.

At this stage I devoted myself especially to the purification of the radium, my husband being absorbed by the study of the physical properties of the rays emitted by the new substances. It was only after treating one ton of pitchblende residues that I could get definite results.

Indeed we know to-day that even in the best minerals there are not more than a few decigrammes of radium in a ton of raw material.

At last the time came when the isolated substances showed all the characters of a pure chemical body. This body, the radium, gives a characteristic spectrum, and I was able to determine for it an atomic weight much higher than that of the barium. This was achieved in 1902. I then possessed one decigramme of very pure radium chloride. It had taken me almost four years to produce the kind of evidence which chemical science demands, that radium is truly a new element. One year would probably have been enough for the same purpose, if reasonable means had been at my disposal. The demonstration that cost so much effort was the basis of the new science of radioactivity.

In later years I was able to prepare several decigrammes of pure radium salt, to make a more accurate determination of the atomic weight and even to isolate the pure radium metal. However, 1902 was the year in which the existence and character of radium were definitely established.

We had been able to live for several years entirely engrossed in the work of research, but gradually circumstances changed. In 1900 my husband was offered a professorship in the University of Geneva, but almost simultaneously he obtained a position of assistant professor at the Sorbonne, and I was made professor at the Normal Superior School for young girls at Sèvres. So we remained in Paris.

I became much interested in my work in the Normal School, and endeavored to develop more fully the practical laboratory exercises of the pupils. These pupils were girls of about twenty years who had entered the school after severe examination and had still to work very seriously to meet the requirements that would enable them to be named professors in the lycées. All these young women worked with great eagerness, and it was a pleasure for me to direct their studies in physics.

But a growing notoriety, because of the announcement of our discoveries, began to trouble our quiet work in the laboratory, and, little by little, life became more difficult. In 1903 I finished my doctor's thesis and obtained the degree. At the end of the same year the Nobel prize was awarded jointly to Becquerel, my husband and me for the discovery of radioactivity and new radioactive elements.

This event greatly increased the publicity of our work. For some time there was no more peace. Visitors and demands for lectures and articles interrupted every day.

The award of the Nobel prize was a great honor. It is also known that the material means provided by this prize was much greater than is usual in prizes for science. This was a great help in the continuation of our researches. Unhappily, we were overtired and had a succession of failures of health for the one or the other of us, so that it was not until 1905 that we were able to go to Stockholm, where my husband gave his Nobel lecture and where we were well received.

The fatigue resulting from the effort exceeding our forces, imposed by the unsatisfactory conditions of our labor, was augmented by the invasion of publicity. The overturn of our voluntary isolation was a cause of real suffering for us and had all the effect of disaster. It was serious trouble brought into the organization of our life, and I have already explained how indispensable was our freedom from external distraction, in order to maintain our family life and our scientific activity. Of course, people who contribute to that kind of trouble generally mean it kindly. It is only that they do not realize the conditions of the problem.

In 1904 our second daughter. Eve Denise, came to us. I had, of course, to interrupt my work in the laboratory for a while. In the same year, because of the awarding of the Nobel prize and the general public recognition, a new chair of physics was created in Sorbonne, and my husband was named as its occupant. At the same time I was named chief of work in the laboratory that was to be created for him. But in reality the laboratory was not constructed then, and only a few rooms taken from other uses were available to us.

In 1906 just as we were definitely giving up the old shed laboratory where we had been so happy, there came the dreadful catastrophe which took my husband away from me and left me alone to bring up our children and, at the same time, to continue our work of research.

It is impossible for me to express the profoundness and importance of the crisis brought into my life by the loss of the one who had been my closest companion and best friend. Crushed by the blow, I did not feel able to face the future. I could not forget, however, what my husband used sometimes to say, that, even deprived of him, I ought to continue my work.

The death of my husband, coming immediately after the general knowledge of the discoveries with which his name is associated, was felt by the public, and especially by the scientific circles, to be a national misfortune. It was largely under the influence of this emotion that the Faculty of Sciences of Paris decided to offer me the chair, as professor, which my husband had occupied only one year and a half in the Sorbonne. It was an exceptional decision, as up to then no woman had held such a position. The University by doing this offered me a precious mark of esteem and gave me opportunity to pursue the researches which otherwise might have had to be abandoned. I had not expected a gift of this kind; I never had any other ambition than to be able to work freely for science. The honor that now came to me was deeply painful under the cruel circumstances of its coming. Besides I wondered whether I would be able to face such a grave responsibility. After much hesitation, I decided that I ought at least to try to meet the task, and so I began in 1906 my teaching in the Sorbonne, as assistant professor, and two years later I was named titular professor.

In my new situation the difficulties of my life were considerably augmented, as I alone had now to carry the burden formerly weighing on my husband and me together. The cares of my young children required close vigilance; in this, my husband's father, who continued to live with us, willingly took his share. He was happy to be occupied with the little girls, whose company was his chief consolation after his son's death. By his effort and mine, the children had a bright home, even if we lived with our inner grief, which they were too young to realize. The strong desire of my father-in-law being to live in the country, we took a house with a garden in Sceaux, a suburb of Paris, from which I could reach the city in half an hour.

This country life had great advantages, not only for my father-in-law, who enjoyed his new surroundings, and especially his garden, but also for my girls, who had the benefit of walks in the open country. But they were more separated from me, and it became necessary to have a governess for them. This position was filled first by one of my cousins, and then by a devoted woman who had already brought up the daughter of one of my sisters. Both of them were Polish, and in this way my daughters learned my native tongue. From time to time, some one of my Polish family came to see me in my grief, and we managed to meet in vacation time, at the seashore in France, and once in the mountains of Poland.

In 1910 we suffered the loss of my very dear father-in-law, after a long illness, which brought me many sorrowful days. I used to spend at his bedside as much time as I could, listening to his remembrances of passed years. His death affected deeply my elder daughter, who, at twelve, knew the value of the cheerful hours spent in his company.

There were few resources for the education of my daughters in Sceaux. The youngest one, a small child, needed principally a hygienic life, outdoor walks and quite elementary schooling. She had already shown a vivid intelligence and an unusual disposition for music. Her elder sister resembled her father in the form of her intelligence. She was not quick, but one could already see that she had a gift of reasoning power and that she would like science. She had some training in a private school in Paris, but I had not wanted to keep her in a lycée, as I have always found the class hours in these schools too long for the health of the children.

My view is that in the education of children the requirement of their growth and physical evolution should be respected, and that some time should be left for their artistic culture. In most schools, as they exist to-day, the time spent in various reading and writing exercises is too great, and the study required to be done at home too much. I also find these schools lacking, in general, in practical exercises to accompany the scientific studies.

With a few friends in the university circle who shared these views, we organized, therefore, a cooperative group for the education of our children, each of us taking charge of the teaching of a particular subject to all of the young people. We were all very busy with other things, and the children varied in age. Nevertheless, the little experiment thus made was very interesting. With a small number of classes we yet succeeded in reuniting the scientific and literary elements of a desirable culture. The courses in science were accompanied by practical exercises in which the children took great interest.

This arrangement, which lasted two years, proved to be very beneficial for most of the children; it was certainly so for my elder daughter. Following this preparation, she was able to enter a higher class in one of the collèges of Paris, and had no difficulty in passing her bachelor's examination before the usual age, after which she continued her scientific studies in the Sorbonne.

My second daughter, although not benefiting by a similar arrangement for her earlier studies, at first followed the classes of a collège only partially, and later completely. She showed herself a good pupil, doing satisfactory work in all directions.

I wanted very much to assure for my children a rational physical education. Next to outdoor walks, I attach a great importance to gymnastics and sports. This side of a girl's education is still rather neglected in France. I took care that my children did gymnastics regularly. I was also careful to have them spend vacations either in the mountains or at the seashore. They can canoe and swim very well and are not afraid of a long walk or a bicycle ride.

But of course the care of my children's education was only a part of my duties, my professional occupations taking most of my time. I have been frequently questioned, especially by women, how I could reconcile family life with a scientific career. Well, it has not been easy; it required a great deal of decision and of self-sacrifice. However, the family bond has been preserved between me and my now grown-up daughters, and life is made brighter by the mutual affection and understanding in our home, where I could not suffer a harsh word or selfish behavior.

In 1906, when I succeeded my husband at the Sorbonne, I had only a provisional laboratory with little space and most limited equipment. A few scientists and students had already been admitted to work there with my husband and me. With their help, I was able to continue the course of research with good success.

In 1907, I received a precious mark of sympathy from Mr. Andrew Carnegie, who donated to my laboratory an annual income for research fellowships which enabled some advanced students or scientists to devote their whole time to investigation. Such foundations are very encouraging to those whose inclinations and talents are such as to warrant their entire devotion to research work. They ought to be multiplied in the interest of science.

As for myself, I had to devote again a great deal of time to the preparation of several decigrammes of very pure radium chloride. With this I achieved, in 1907, a new determination of the atomic weight of radium, and in 1910 I was able to isolate the metal. The operation, an extremely delicate one, was performed with the assistance of a distinguished chemist belonging to the laboratory staff. It has never been repeated since that time, because it involves a serious danger of loss of radium, which can be avoided only with utmost care. So I saw at last the mysterious white metal, but could not keep it in this state, for it was required for further experiments.

As for the polonium, I have not been able to isolate it, its quantity in the mineral being even much less than the quantity of radium. However, very concentrated polonium has been prepared in my laboratory, and important experiments have been performed with this substance, concerning especially the production of helium by radiation of polonium.

I had to devote special care to the improvement of the measuring methods in the laboratory. I have told how important precise measurements were in the discovery of radium. It is still to be hoped that efficient methods of quantitative determination may lead to new discoveries.

I devised a very satisfactory method for determining the quantity of radium by the means of a radioactive gas produced by it and called "emanation." This method, frequently used in my laboratory, permits of the measurement of very small quantities of radium (less than a thousandth of a milligramme), with a fair precision. More important quantities are often measured by their penetrating radiation, named Gamma-rays. For this we also possess in my laboratory a suitable equipment. It is easier and more satisfactory to measure the radium by the emitted rays, than to weigh it in a balance. However, these measurements require the disposition of reliable standards. So the question of a radium standard had to be taken into careful consideration.

The measurements of radium had to be established on a solid basis, for the benefit of laboratories and scientific research, which, of course, is in itself an important reason, and moreover, the growing medical utilization of this substance made it necessary to control the relative purity of commercially produced radium.

The first experiments on the biological properties of radium were successfully made in France with samples from our laboratory, while my husband was living. The results were, at once, encouraging, so that the new branch of medical science, called radiumtherapy (in France, Curietherapy), developed rapidly, first in France and later in other countries. To supply the radium wanted for this purpose, a radium-producing industry was established. The first plant was created in France and worked very successfully, but afterwards manufactures were founded in other countries, the most important of which are now in America, where great quantities of radium ore, named "carnotite," are available. The radiumtherapy and the radium production developed conjointly, and the results were more and more important, for the treatment of several diseases, and particularly of cancer. As a consequence of this, several institutes have been founded, in the large cities, for the application of the new therapy. Some of these institutes own several grammes of radium, the commercial price of the gramme being now about $70,000, the cost of production depending on the very small proportion of radium in the ore.

It may be easily understood how deeply I appreciated the privilege of realizing that our discovery had become a benefit to mankind, not only through its great scientific importance, but also by its power of efficient action against human suffering and terrible disease. This was indeed a splendid reward for our years of hard toil.

The success of the therapy depends, of course, on the precise knowledge of the quantity of radium which is used, so that the measurements of radium are as important for industry and for medicine as for physicochemical research.

Considering all these needs, a commission of scientific men of different countries was formed who agreed to take as a base an international standard, formed of a carefully weighed quantity of pure radium salt. Secondary standards were then to be prepared for each country, and compared to the basic standard by means of their radiation. I was appointed to prepare the primary standard.

This was a very delicate operation, as the weight of the standard sample, quite small (about 21 milligrammes of chloride), had to be determined with great precision. I performed the preparation in 1911. The standard is a thin glass tube, of a few centimeters in length, containing the pure salt which was used for the determination of atomic weight. It was accepted by the Commission and is deposited in the International Bureau of Weights and Measures at Sèvres, near Paris. Several secondary standards, compared with the primary one, have been put into service by the Commission. In France the control of radium tubes, by the measurement of their radiation, takes place in my laboratory, where any one may bring the radium to be tested; in the United States this is done in the Bureau of Standards.

Near the end of the year 1910, I was proposed for the decoration of the Légion of Honor. A similar proposal was made earlier in favor of my husband, who, however, being opposed to all honorary distinctions, did not accept the nomination. As my husband and I were too united in all things for me to act differently from him in this matter, I did not accept the decoration, in spite of the insistence of the Ministry. At that time also, several colleagues persuaded me to be a candidate for election to the Academy of Sciences of Paris, of which my husband was a member during the last months of his life. I hesitated very much, as such a candidacy requires, by custom, a great number of personal visits to Academy members. However, I consented to offer myself a candidate, because of the advantages an election would have for my laboratory. My candidacy provoked a vivid public interest, especially because it involved the question of the admission of women to the Academy. Many of the Academicians were opposed to this in principle, and when the scrutiny was made, I had a few votes less than was necessary. I do not ever wish to renew my candidacy, because of my strong distaste for the personal solicitation required. I believe that all such elections should be based wholly on a spontaneous decision, without any personal efforts involved, as was the case for several Academies and Societies which made me a member without any demand or initiative on my part.

As a result of all the cares devolving on me, I fell seriously ill at the end of 1911, when, for the second time, I received, this time alone, the award of the Nobel prize. This was a very exceptional honor, a high recognition of the discovery of the new elements and of the preparation of pure radium. Suffering though I was, I went to Stockholm to receive the prize. The journey was extremely painful for me. I was accompanied by my eldest sister and my young daughter Irene. The ceremony of delivery of the Nobel prizes is very impressive, having the features of a national solemnity. A most generous reception was accorded me, specially by the women of Sweden. This was a great comfort to me, but I was suffering so much that when I returned I had to stay in bed for several months. This grave illness, as well as the necessities of my children's education, obliged me to move my home from Sceaux to Paris.

During the year 1912 I had the opportunity of collaborating in the creation of a laboratory of radium at Warsaw. This laboratory was founded by the Scientific Society of Warsaw which offered me its direction. I could not leave France to go back to my native country, but I willingly agreed to occupy myself with the organization of the studies in the new laboratory. In 1913, having improved my health, I was able to attend an inauguration fête in Warsaw, where a touching reception was given, leaving me an unforgettable memory of national sentiment which succeeded in creating useful work under particularly difficult political conditions.

While still only partially recovered from my illness, I renewed my efforts for the construction of a suitable laboratory in Paris. Finally it was arranged for, and work began in 1912. The Pasteur Institute wished to be associated with this laboratory, and, in accord with the University, it was decided to create an Institute of Radium, with two laboratories, one of physics and one of biology, the first to be devoted to studies of the physical and chemical properties of the radioactive elements, the second to the study of their biological and medical applications. But, because of the lack of financial means, the construction work proceeded very slowly, and was not yet entirely finished when the war broke out in 1914.

# CHAPTER III

In 1914, it happened, as it often had in other years, that my daughters had left Paris for their summer vacation before me. They were accompanied by their governess, in whom I had all confidence, and were living in a small house on the seashore in Brittany, at a place where there were also the families of several of our good friends. My work did not generally permit me to pass the entire vacation near them without interruption.

That year I was preparing to join them in the last days of July, when I was stopped by the bad political news, with its premonitions of an imminent military mobilization. It did not seem possible for me to leave under these conditions, and I waited for further events. The mobilization was announced on August 1st, immediately followed by Germany's declaration of war on France. The few men of the laboratory staff and the students were mobilized, and I was left alone with our mechanic who could not join the army because of a serious heart trouble.

The historic events that followed are known to every one, but only those who lived in Paris through the days of August and September, 1914, can ever really know the state of mind in the capital and the quiet courage shown by it. The mobilization was a general wave of all France passing out to the border for the defense of the land. All our interest now centered on the news from the front.

After the uncertainties of the first days this news became more and more grave.

First, it was the invasion of Belgium and the heroic resistance of that little country; then the victorious march of the German army through the valley of the Oise toward Paris; and soon the departure of the French government to Bordeaux, followed by the leaving of those Parisians who could not, or would not, face the possible danger of German occupation. The overloaded trains took into the country a great number of people, mostly of the well-to-do class. But, on the whole, the people of Paris gave a strong impression of calm and quiet decision in that fateful year of 1914. In the end of August and the beginning of September the weather was radiant, and under the glorious sky of those days the great city with its architectural treasures seemed to be particularly dear to those who remained in it.

When the danger of German attack on Paris became pressing, I felt obliged to put in security the supply of radium then in my laboratory, and I was charged by the government to take it to Bordeaux for safety. But I did not want to be away long, and hence decided to return immediately. I left by one of the trains that were carrying government staff and baggage, and I well remember the aspect of the national highway which is at intervals in view from the train; it showed a long line of motor-cars carrying their owners from the capital.

Arriving at Bordeaux in the evening, I was very embarrassed with my heavy bag including the radium protected by lead. I was not able to carry it and waited in a public place, while a friendly ministry employee who came by the same train managed to find a room for me in a private apartment, the hotels being overcrowded. The next morning I hurried to put the radium in a safe place, and succeeded, although not without difficulty, in taking a military train back to Paris in the evening of the same day. Having opportunity for exchanging a few sentences with persons on the place who wanted to ask information from people coming by the train, I was interested to notice how they seemed surprised and comforted to learn of some one who found it natural to return to Paris.

My trip back was troubled by delays; for several hours the train rested immovable on the rails, while the travelers accepted a little bread from the soldiers who were provided with it. Finally arriving in Paris, I learned that the German army had turned; the battle of the Marne had begun.

In Paris I shared the alternating hope and grief of the inhabitants during the course of that great battle, and had the constant worry of foreseeing a long separation from my children in case the Germans succeeded in occupying the city. Yet I felt that I must stay at my post. After the successful outcome of the battle, however, any immediate danger of occupation being removed, I was able to have my daughters come back from Brittany to Paris and again take up their studies. This was the great desire of my children, who did not want to stay away from me and from their work, even if many other families thought it wiser to stay in the country, far from the front.

The dominant duty imposed on every one at that time was to help the country in whatever way possible during the extreme crisis that it faced. No general instructions to this were given to the members of the University. It was left to each to take his own initiative and means of action. I therefore sought to discover the most efficient way to do useful work, turning my scientific knowledge to most profit.

During the rapid succession of events in August, 1914, it was clearly proved that the preparation for defense was insufficient. Public feeling was especially aroused by the realization of the grave failings which appeared in the organization of the Health Service. My own attention was particularly drawn to this situation, and I soon found a field of activity which, once entered upon, absorbed the greatest part of my time and efforts until the end of the war, and even for some time thereafter. The work was the organization of radiologic and radiotherapeutic services for the military hospitals. But I also had to make the change, during these difficult war years, of my laboratory into the new building of the Institute of Radium and to continue, in the measure possible to me, regular teaching, as well as to investigate certain problems especially interesting the military service.

It is well known that the X-rays offer surgeons and doctors extremely useful means for the examination of the sick and wounded. They make possible the discovery and the exact location of projectiles which have entered the body, and this is a great help in their extraction. These rays also reveal lesions of bones and of the internal organs and permit one to follow the progress of recovery from internal injuries. The use of the X-rays during the war saved the lives of many wounded men; it also saved many from long suffering and lasting infirmity. To all the wounded it gave a greater chance of recovery.

However, at the beginning of the war, the Military Board of Health had no organization of radiology, while the civil organization was also but little developed. Radiologic installations existed in only a small number of important hospitals, and there were only a few specialists in the large cities. The numerous new hospitals that were established all over France in the first months of the war had, as a rule, no installation for the use of X-rays.

To meet this need I first gathered together all the apparatus I could find in the laboratories and stores. With this equipment I established in August and September, 1914, several stations of radiology, the operation of which was assured by volunteer helpers to whom I gave instruction. These stations rendered great service during the battle of the Marne. But as they could not satisfy the needs of all the hospitals of the Paris region, I fitted up, with the help of the Red Cross, a radiologic car. It was simply a touring motor-car, arranged for the transport of a complete radiologic apparatus, together with a dynamo that was worked by the engine of the car, and furnished the electric current necessary for the production of the rays. This car could come at the call of any of the hospitals, large or small, in the surroundings of Paris. Cases of urgent need were frequent, for these hospitals had to take care of the wounded who could not be transported to more distant places.

The first results of this work showed that it was necessary to do more. Thanks to special donations and to the help of a very efficient relief committee called "le Patronage National des Blessés," I succeeded in developing my initiative to a considerable extent. About two hundred radiologic installations were established or materially improved through my efforts in the zone of the French and Belgian armies, and in the regions of France not occupied by the army. I was able, besides, to equip in my laboratory and give to the army twenty radiologic cars. The frames of these cars were donated by various persons who wished to be helpful; some of them offered also the equipment. The cars were of the greatest service to the army.

These privately developed installations were particularly important in the first two years of the war, when the regular military service possessed but few radiologic instruments. Later the Board of Health created, little by little, a considerable radiologic service of its own, as the utility of the stations was more clearly realized owing to the example given by private initiative. But the needs of the armies were so great, that my cooperation continued necessary to the end of the war, and even afterwards.

I could not have accomplished this work without seeing for myself the needs of the ambulance stations and hospitals. Thanks to the help of the Red Cross and to the agreement of the Board of Health, I was able to make several journeys to the army zones and to the other parts of France. Several times I visited the ambulance stations of the armies of the north and in the Belgian zone, going to Amiens, Calais, Dunkirk, Furnes, and Poperinghe. I went to Verdun, Nancy, Luneville, Belfort, to Compiegne, and Villers-Cotterets. In the regions distant from the front, I took care of many hospitals which had to do very intensive work with little aid. And I keep as a precious recollection of that time, many letters of warm recognition from those to whom I brought help in their difficulties.

The motive of my starting on a journey was usually a demand from surgeons. I went with a radiologic car which I kept for my personal use. In examining the wounded in the hospital, I could gain information of the special needs of the region. When back in Paris, I got the necessary equipment to meet these needs and returned to install it myself, for very often the people on the ground could not do it. I had then to find competent persons to handle the apparatus and show them how to do it, in full detail. After a few days of hard toil, the manipulator knew enough to work the apparatus himself, and at the same time a large number of wounded had been examined. In addition, the surgeons of the region had gained an idea of the usefulness of the radiologic examination (which few of them knew at that time), and friendly relations were established which made the later development of my work much easier.

On several of my trips I was accompanied by my elder daughter, Irene, who was then seventeen years old, and, having finished her preparatory studies, was beginning higher studies at the Sorbonne. Because she greatly desired to be useful, she now studied nursing and learned radiology, and did her best to help me under the most varied circumstances. She did ambulance work at the front between Furnes and Ypres, and also at Amiens, receiving, from the Chiefs of Service, testimonials of work satisfactorily performed and, at the end of the war, a medal.

Of the hospital life of those years, we keep many a remembrance, my daughter and I. Traveling conditions were extraordinarily difficult; we were often not sure of being able to press forward, to say nothing of the uncertainty of finding lodgings and food. However, things always ended in arranging themselves, thanks to our persistence and to the good will we met. Wherever we went I had to look after each detail myself and see innumerable military chiefs to obtain passes and permissions for transportation. Many a time I loaded my apparatus on to the train myself, with the help of the employees, to make sure that it would go forward instead of remaining behind several days at the station. And on arrival I also went to extract them from the encumbered station.

When I traveled with the radiologic car, other problems presented themselves. I had, for instance, to find safe places for the car, to get lodgings for the assistants and to secure the automobile accessories. Since chauffeurs were scarce, I learned to drive the car, and did it when necessary. Owing to all this personal supervision, my installations were usually swiftly made, whereas appeal to the Central Health Service was answered slowly. So the military chiefs greatly appreciated the assistance they could get from me, especially in cases of urgent need.

We both, my daughter and myself, have pleasant and grateful memories of the personnel of the hospitals, and were on the best terms with the surgeons and nurses. One could not but admire these men and women who were giving their services without counting, and whose task was often overwhelming. Our collaboration was easy, for my daughter and I tried to work in their spirit; and we felt that we were standing side by side with friends.

While we were attached to the Belgian Ambulance Service, we were present several times during visits of King Albert and Queen Elizabeth. We appreciated deeply their devotion, their solicitude for the wounded, their extreme simplicity, and the cordiality of their behavior.

But nothing was so moving as to be with the wounded and to take care of them. We were drawn to them because of their suffering and because of the patience with which they bore it. Almost everyone did his best to facilitate the X-ray examination, notwithstanding the pain caused by any displacement. One learned very soon to know them individually and to exchange with them a few friendly words. Those who were not familiar with the examination, wanted very much to be reassured about the effect of the strange apparatus they were going to experience.

I can never forget the terrible impression of all that destruction of human life and health. To hate the very idea of war, it ought to be sufficient to see once what I have seen so many times, all through those years: men and boys brought to the advanced ambulance in a mixture of mud and blood, many of them dying of their injuries, many others recovering but slowly through months of pain and suffering.

One of my difficult problems was to find the necessary trained assistants to operate my apparatus. At the beginning of the war there was little knowledge of radiology, and apparatus in the hands of those who did not understand how to handle it deteriorated quickly and was soon useless. The practice of radiology in most hospitals in war-time does not require much medical knowledge; it can be sufficiently grasped by intelligent persons who know how to study and who have some notion of electrical machinery. Professors, engineers, or university students often made good manipulators. I had to look for those who were temporarily free from military service, or who happened to be stationed in the locality where I needed them. But even after I had secured them, these operators were often transferred by military orders, and I had to search again for others to fill their places. For this reason, I determined to train women to do this work.

Accordingly, I proposed to the Health Service to add a department of radiology to the Nurses' School which had just been founded at the Edith Cavell Hospital. This they agreed to do. And so, in 1916, the course was organized at the Radium Institute, and provided in the following years of war for the training of one hundred and fifty operators. Most of the pupils who applied had only an elementary education, but could succeed if working in a proper way. The course comprised theoretical studies and very extended practical training; it included also some instruction in anatomy. It was given by a few persons of good will, among them my daughter. Our graduates formed an excellent personnel very genuinely appreciated by the Board of Health. Theoretically, they were supposed to serve as aides to physicians, but several of them proved capable of independent work.

My continued and various experience in war radiology gave me a wide knowledge of that subject, which I felt should be made more familiar to the public. So I wrote a small book called "Radiology and the War," in which I aimed to demonstrate the vital importance of radiology and to compare its development during war time with its use in the previous time of peace.

I come now to the account of the founding of the service of radiumtherapy at the Radium Institute.

In 1915, the radium, which had been safely deposited in Bordeaux, was brought back to Paris, and not having time for regular scientific research, I decided to use it to cure the wounded, without, however, risking the loss of this precious material. I proceeded to place at the disposal of the Health Service not the radium itself, but the emanation which can be obtained from it at regular intervals. The technique of the use of the emanation can readily be employed in the larger radiumtherapy institutes, and, in many ways, is more practicable than the direct use of radium. In France, however, there was no national institute of radiumtherapy, and the emanation was not used in hospitals.

I offered to furnish regularly to the Health Service bulbs of radium emanation. The offer was accepted, and the "Emanation Service," started in 1916, was continued until the end of the war and even longer. Having no assistants, I had, for a long time, to prepare these emanation bulbs alone, and their preparation is very delicate. Numbers of wounded and sick, military and civil, were treated by means of these bulbs.

During the bombardment of Paris, the Health Board took special measures to protect from shells the laboratory in which the bulbs were prepared. Since the handling of radium is far from being free of danger (several times I have felt a discomfort which I consider a result of this cause), measures were taken to prevent harmful effects of the rays on the persons preparing emanation.

While the work in connection with the hospitals remained my major interest, I had many other preoccupations during the war.

After the failure of the German offensive in the summer of 1918, at the request of the Italian government, I went to Italy to study the question of her natural resources in radioactive materials. I remained a month and was able to obtain certain results in interesting the public authorities in the importance of this new subject.

It was in 1915 that I had to move my laboratory to the new building in the rue Pierre Curie. This was a trying and complicated experience, for which, once more, I had no money nor any help. So it was only between my journeys that I was able, little by little, to do the transportation of my laboratory equipment, in my radiologic cars. Afterwards, I had much work in classifying and distributing my materials, and arranging the new place in general, with the help of my daughter and of my mechanic, who, unfortunately, was often ill.

One of my first cares was to have trees planted in the limited grounds of my laboratory. I feel it very necessary for the eyes to have the comfort of fresh leaves in spring and summer time. So I tried to make things pleasant for those who were to work in the new building. We planted a few lime trees and plane trees, as many as there was room for, and did not forget flowerbeds and roses. I well remember the first day of bombardment of Paris with the big German gun; we had gone, in the early morning, to the flower-market, and spent all that day busy with our plantation, while a few shells fell in the vicinity.

In spite of the great difficulties, the new laboratory was organized little by little, and I had the satisfaction of having it quite ready for the beginning of the school-year 1919-20, the period of demobilization. In the spring of 1919, I organized special courses for some American soldier students, who also studied with much zeal the practical exercises directed by my daughter.

Mme. Curie instructing American soldiers in her Paris laboratory. Photo by U. S. Signal Corps.

The entire period of the war was for me, as for many others, a period of great fatigue. I took almost no vacation, except for a few days, now and then, when I went to see my daughters on their holidays. My older daughter would scarcely take any, and I was obliged to send her away sometimes to preserve her health. She was continuing her studies in the Sorbonne, and besides, as said before, was helping me with my war work, while the younger daughter was still in the preparatory college. Neither of them wished to leave Paris during the bombardment.

After more than four years of a war which caused ravages without precedent, the armistice came at last, in the autumn of 1918, followed by laborious efforts to reëstablish peace, which is not yet general nor complete. It was a great relief to France to see the end of that dark period of cruel losses. But the griefs are too recent and life still too hard for calm and happiness yet to be restored.

Nevertheless, a great joy came to me as a consequence of the victory obtained by the sacrifice of so many human lives. I had lived, though I had scarcely expected it, to see the reparation of more than a century of injustice that had been done to Poland, my native country, and that had kept her in slavery, her territories and people divided among her enemies. It was a deserved resurrection for the Polish nation, which showed herself faithful to her national memories during the long period of oppression, almost without hope. The dream that appeared so difficult to realize, although so dear, became a reality following the storm that swept over Europe. In these new conditions I went to Warsaw and saw my family again, after many years of separation, in the capital of free Poland. But how difficult are the conditions of life of the new Polish republic, and how complicated is the problem of reorganization after so many years of abnormal life!

In France, partly devastated and suffering from the loss of so many of her citizens, the difficulties created by the war are not yet effaced, and the return to normal work is being attained only gradually. The scientific laboratories feel this state of affairs and the same condition prevails for the Radium Institute.

The various radiologic organizations created during the war still partially exist. The Radiographic Nurses' School has been maintained at the request of the Board of Health. The emanation service, which could not be abandoned, is also continued in a considerably enlarged form. It has passed under the direction of Doctor Regaud, Director of the Pasteur Laboratory of the Radium Institute, and is developing into a great national service of radiumtherapy.

The work of the laboratory has been reorganized, with the return of the mobilized personnel and the students. But in the restrained circumstances under which the country still exists, the laboratory lacks ways and means for its efficient development. Particularly are wanted an independent hospital for radiumtherapy (which is called Curietherapy in France), and an experimental station, outside of Paris, for experiments on great quantities of material, such as are needed for the progress of our knowledge of radioactive elements.

I myself am no longer young, and I frequently ask myself whether, in spite of recent efforts of the government aided by some private donations, I shall ever succeed in building up for those who will come after me an Institute of Radium, such as I wish to the memory of Pierre Curie and to the highest interest of humanity.

However, a precious encouragement came to me in the year 1921. On the initiative of a generous daughter of the United States, Mrs. W. B. Meloney, the women of that great American country collected a fund, the "Marie Curie Radium Fund," and offered me the gift of a gramme of radium to be placed entirely at my disposal for scientific research. Mrs. Meloney invited me with my daughters to come to America and to receive the gift, or the symbol of it, from the hands of the President of the great republic, at the White House.

The fund was collected by a public subscription, as well by small as by important gifts, and I was very thankful to my sisters of America for this genuine proof of their affection. So I started for New York at the beginning of May, after a ceremony given in my honor at the Opera of Paris, to greet me before my departing.

I keep a grateful memory of my sojourn in the United States for several weeks, of the impressive reception at the White House, where President Harding addressed me in generous and affectionate words, of my visits to the universities and colleges which welcomed me and bestowed on me their honorary degrees, of the public reunions where I could not but feel the deep sympathy of those who came to meet me and to wish me good luck.

Madame Curie in her laboratory at the Institut Curie, Paris. Henri Manuel, Paris.

I had also the opportunity of a visit to the Niagara Falls and to the Grand Canyon, and admired immensely these marvelous creations of nature.

Unhappily, the precarious state of my health did not permit of the complete fulfilment of the general plan established by my visit to America. However, I saw and learned much, and my daughters enjoyed to a full extent the opportunities of their unexpected vacation and the pride in the recognition of their mother's work. We left for Europe at the end of June, with the real sorrow of parting from excellent friends whom we would not forget.

I came back to my work, made easier by the precious gift, with an even stronger desire to carry it forward with renewed courage. But as my aims are still wanting support in essential parts, I am frequently compelled to give thought to a very fundamental question concerning the view a scientist ought to take of his discovery.

My husband, as well as myself, always refused to draw from our discovery any material profit. We have published, since the beginning, without any reserve, the process that we used to prepare the radium. We took out no patent and we did not reserve any advantage in any industrial exploitation. No detail was kept secret, and it is due to the information we gave in our publications that the industry of radium has been rapidly developed. Up to the present time this industry hardly uses any methods except those established by us. The treatment of the minerals and the fractional crystallizations are still performed in the same way, as I did it in my laboratory, even if the material means are increased.

As for the radium prepared by me out of the ore we managed to obtain in the first years of our work, I have given it all to my laboratory.

The price of radium is very high since it is found in minerals in very small quantities, and the profits of its manufacture have been great, as this substance is used to cure a number of diseases. So it is a fortune which we have sacrificed in renouncing the exploitation of our discovery, a fortune that could, after us, have gone to our children. But what is even more to be considered is the objection of many of our friends, who have argued, not without reason, that if we had guaranteed our rights, we could have had the financial means of founding a satisfactory Institute of Radium, without experiencing any of the difficulties that have been such a handicap to both of us, and are still a handicap to me. Yet, I still believe that we have done right.

Humanity, surely, needs practical men who make the best of their work for the sake of their own interests, without forgetting the general interest. But it also needs dreamers, for whom the unselfish following of a purpose is so imperative that it becomes impossible for them to devote much attention to their own material benefit. No doubt it could be said that these idealists do not deserve riches since they do not have the desire for them. It seems, however, that a society well organized ought to assure to these workers the means for efficient labor, in a life from which material care is excluded so that this life may be freely devoted to the service of scientific research.

# CHAPTER IV

## A VISIT TO AMERICA

My beautiful voyage to the United States of America resulted, as is known, from the generous initiative of an American woman, Mrs. Meloney, editor of an important magazine, the Delineator, who, having planned the gift of a gramme of radium to me by her countrywomen, succeeded in a few months in bringing this plan to execution, and asked me to come over and receive the gift personally.

The idea was that the gift would come exclusively from the American women. A committee including several prominent women and distinguished scientific men received some important gifts, and made an appeal for a public subscription, to which a great number of women's organizations, especially colleges and clubs, responded. In many cases gifts came from persons who had experienced the benefit of radiumtherapy. In this way was collected the "Marie Curie Radium Fund" of more than one hundred thousand dollars for the purchase of a gramme of radium. The President of the United States, Mr. Harding, kindly agreed to deliver the gift in a ceremony at the White House.

The Committee invited me and my daughters to the United States in May, and even though it was not vacation time for me, I accepted the invitation with the consent of the University of Paris.

All care of the voyage was taken away from me. Mrs. Meloney came to France in time to be present at a manifestation organized on the 28th of April in favor of the Radium Institute of Paris by the magazine Je Sais Tout, and accompanied by sincere expressions of sympathy for the American nation. On May 4th, we took passage at Cherbourg on the Olympic for New York.

The program of my voyage prepared by the Committee seemed very intimidating. It was announced that I would not only attend the ceremony at the White House, but also visit many universities and colleges in several towns. Some of these institutions had contributed to the Fund; all desired to offer me honors. The vitality and the activity of the American nation produces programs on a large scale. On the other hand, the wideness of the country has developed in American citizens the custom of long travel. But during all that travel I was protected with the greatest care, in order to lighten as far as possible the inevitable fatigue of the voyage and the receptions. America not only gave me a generous welcome, but also true friends whom I could not thank enough for their kindness and their devotion.

After having admired the grand view of the harbors of New York, and having been greeted by groups of students, Girl Scouts, and Polish delegates, and welcomed by many gifts of flowers, we took possession of a peaceful apartment in town. The following day I made the acquaintance of the Reception Committee at a luncheon given by Mrs. Carnegie in her beautiful home still filled with memories of her husband, Andrew Carnegie, whose philanthropic achievements are well known in France. The following day we went for a visit of a few days to Smith College, and Vassar College, a few hours from New York. Later I also visited the colleges of Bryn Mawr and Wellesley, and I saw some others on my way.

These colleges, or universities for women, are very characteristic of American life and culture. My short visit could not permit me to give an authorized opinion on the intellectual training, but even in such a visit as I made one may notice important differences between the French and American conception of girls' education, and some of these differences would not be in favor of our country. Two points have particularly drawn my attention: the care of the health and the physical development of the students, and the very independent organization of their life which allows a large degree of individual initiative.

The colleges are excellent in their construction and organization. They are composed of several buildings, often scattered in very large grounds between lawns and trees. Smith is on the shore of a charming river. The equipment is comfortable and hygienic, of extreme cleanliness, with bathrooms, showers, distribution of cold and hot water. The students have cheerful private rooms and common gathering rooms. A very complete organization of games and sports exists in every college. The students play tennis and baseball; they have gymnasium, canoeing, swimming, and horseback riding. Their health is under the constant care of medical advisers. It seems to be a frequent opinion of American mothers that the existing atmosphere of cities like New York is not favorable to the education of young girls, and that a life in the country in the open air gives more suitable conditions for the health and the tranquillity of studying.

In every college the young girls form an association and elect a committee which has to establish the internal rules of the college. The students display a great activity: they take part in educational work; they publish a paper; they are devoted to songs and music, they write plays, and act them in college and out of it. These plays have interested me very much in their subjects and the execution. The students are also of different social conditions. Many of them are of wealthy families, but many others live on scholarships. The whole organization may be considered as democratic. A few students are foreigners, and we have met some French students very well pleased with the college life and the studies.

Every college takes four years of study with examinations from time to time. Some students afterwards do personal work, and acquire the degree of Doctor, which does not exactly correspond to the same title in France. The colleges have laboratories with many good facilities for experimentation.

I have been strongly impressed by the joy of life animating these young girls and expanding on every occasion, like that of one of my visit. If the ceremonies of the reception were performed in a nearly military order, a spontaneity of youth and happiness expressed itself in the songs of greeting composed by the students, in the smiling and excited faces, and in the rushing over the lawns to greet me at my arrival. This was indeed a charming impression which I could not forget.

Back in New York, several ceremonies awaited me before my leaving for Washington. A luncheon of the Chemists, a reception at the

Museum of Natural History and the Mineralogical Club, a dinner at the Institute of Social Sciences, and a great meeting at Carnegie Hall, where many delegations represented the faculties and students of women's colleges and universities. At all these receptions I was greeted in warm addresses by prominent men and women, and I received honors very precious to me because of the sincerity of the givers. Neither has the part of national friendships been forgotten; the address of Vice-President Coolidge was a noble recognition of the past where French and Polish citizens have been helpful to the young American Republic, and is also a statement of fraternity strengthened by the tempest of the last years.

It was in this atmosphere of affection created by the convergence of intellectual and social sympathies that there took place on May 20th the beautiful ceremony at the White House. It was a deeply moving ceremony in all its simplicity, occurring before a democratic gathering including the President and Mrs. Harding, cabinet officers, Judges of the Supreme Court, high officers of Army and Navy, foreign diplomats, representatives of women's clubs and societies, and prominent citizens of Washington and other cities. It comprised a short presentation by the French ambassador, M. Jusserand, a speech by Mrs. Meloney for the American women, the address of President Harding, a few words of gratitude said by me, a defile of the guests, and a group picture for a souvenir, all this in the admirable setting of the White House, peaceful and dignified, white indeed between its green lawns with wide prospects on that beautiful afternoon of May. A remembrance never to be forgotten was left by this reception in which the chief representative of a great nation offered me homage of infinite value, the testimonial of the recognition of his country's citizens.

The address of the President had been inspired by the same sentiments as that of Vice-President Coolidge, as far as concerned his appreciation of France and Poland. This address gave also an expression of the American feeling which was emphasized by an exceptional solemnity in the delivering of the gift.

Mme. Curie and President Harding at the White House, May 20, 1921, when a gram of radium was presented to its discoverer by the women of America.

Copyright International.

The American nation is generous, and always ready to appreciate an action inspired by considerations of general interest. If the discovery of radium has so much sympathy in America, it is not only because of its scientific value, and of the importance of medical utilization; it is also because the discovery has been given to humanity without reservation or material benefits to the discoverers. Our American friends wanted to honor this spirit animating the French science.

The radium itself was not brought to the ceremony. The President presented me with the symbol of the gift, a small golden key opening the casket devised for the transportation of the radium.

Our sojourn at Washington following the principal ceremony included a very agreeable reception at the French Embassy and the Polish Legation, a reception at the National Museum, and some laboratory visits.

The itinerary of our journey from Washington included visits to the cities of Philadelphia, Pittsburgh, Chicago, Buffalo, Boston, and New Haven, a visit to the Grand Canyon, and to Niagara Falls. On that trip I was the guest of several universities which did me the honor of bestowing honorary degrees on me. I have to thank for these the universities of Pennsylvania, of Pittsburgh, of Chicago, the Northwestern University, Columbia University, Yale University, the Women's Medical College of Pennsylvania, the University of Pennsylvania, Smith College, and Wellesley College, while I thank Harvard University for her reception.

The delivery of honorary degrees in American universities is accompanied by solemnities. In principle, the presence of the candidate is required, and the delivery takes place at the annual commencement, but, in some cases, special ceremonies were organized in my favor. The university ceremonies in America are more frequent than in France, and play a more important part in the university life. Especially is this true at the annual commencement, which begins with an academic procession over the grounds of the university, the procession including the officials, the professors, and graduates in academic caps and gowns. Afterwards all assemble in a hall where are announced the diplomas corresponding to the grades of bachelor, master, and doctor. There is always a musical part in the program, and addresses are delivered by the officials of the university or invited orators. These addresses are naturally devoted to dignifying the ideals and the humanitarian purposes of education; but in certain cases it seems permitted to introduced point of American humor. These ceremonies are on the whole very impressive, and certainly contribute to keep a bond between the university and the alumni. This is a favorable circumstance for those great American universities which are sustained entirely on private foundations. It is only in more recent times that most States have created universities supported by the State.

At Yale University I had the pleasure of representing the University of Paris at the inauguration of President Angell, fourteenth president of the University. I was also pleased to attend at Philadelphia a meeting of the American Philosophical Society and a meeting of the College of Physicians, and at Chicago a meeting of the American Chemical Society at which I delivered a lecture on the Discovery of Radium. The medals of John Scott, Benjamin Franklin, and Willard Gibbs have been presented to me by these societies.

Several meetings organized in my honor by the American women's organizations have particularly interested the American public. I have already mentioned the meeting of the University Women at Carnegie Hall of New York; a similar meeting was held at Chicago, where I was also received by the Association of Polish Women. I was also greeted by women's organizations in the Carnegie Institute of Pittsburgh, and by a delegation of Canadian university women at Buffalo. In all these meetings it was impossible not to recognize the sincerity of the emotion in the women who gave me their best wishes, at the same time expressing their confidence in the future of feminine intelligence and activity. I did not feel any opposition between these feministic aspirations and the masculine opinion. As far as I could notice, the men in America approve of these aspirations and encourage them. This is a very favorable condition for the social activity of the American women which reveals itself in a strong interest in work for education, for hygiene, and for the improvement of conditions of labor. But any other unselfish purpose may rely on their support, as is proved by the success of Mrs. Meloney's plan, and by the sympathy this plan encountered in women of all social conditions.

I could not, to my deep regret, give time enough to the visit to laboratories and scientific institutes. These too brief visits were of great interest to me. I found everywhere the greatest care for developing scientific activity and for improving the facilities. New laboratories are in building, and in older laboratories very modern equipment may be found. The available room never gives that impression of insufficiency from which we suffer too often in France. The means are provided by private initiative expressed in gifts and foundations of various kinds. There exists also a National Council of Research established by private funds for stimulating and improving scientific work, and for assuring its connection with industry.

I have visited with special interest the Bureau of Standards, a very important national institution at Washington for scientific measurements and for study connected with them. The tubes of radium presented to me were at the Bureau, whose officials had kindly offered to make the measurements, and to take care of the packing and delivery to the ship.

A new laboratory has been created at Washington for researches on very low temperatures with the use of liquid hydrogen and liquid helium. I had the honor of dedicating this laboratory to its service.

I had the great pleasure of meeting in their laboratories several very important American scientific men. The hours I spent in their company are among the best of my travel.

The United States possesses several hospitals for radiumtherapy. These hospitals are generally provided with laboratories for the extraction of radium emanation which is sealed up in small tubes for medical use. These institutions own important quantities of radium, have a very good equipment, and treat a great number of patients. I have visited some of them, and this made me feel more deeply, if possible, the regret of not having in France even one national institute capable of rendering the same services. I hope that this lack will be filled in the near future.

The industry of radium has been started in France, but it is in America that it has had its greatest development, owing to the presence of a sufficient supply of the ore carnotite.[15] I was very much interested in my visit to the most important of the factories, and I gladly recognize the spirit of initiative in this undertaking. The factory owns a collection of documentary films which enable one to appreciate the effort made each day in collecting the ore scattered in the immense fields of Colorado, in carrying and concentrating this ore originally very poor in radium. On the other hand, the means of extraction of radium are still the same which have been described in earlier chapters.

The greatest courtesy was paid me in my visit to the radium plant and laboratory. I found the same reception at a factory of mesothorium which presented me with some material, and where the officials expressed the desire to help in my scientific work.

To make complete these travel impressions it would be necessary to speak of the nature of the country. I recoil before the task, being incapable of expressing in a few words the immensity and the variety of the spaces which opened before my eyes. The general impression is one of unlimited possibilities for the future. I keep a particularly vivid remembrance of the great falls of Niagara, and of the magnificent colors of the Grand Canyon.

On June 28th I embarked in New York on the same ship which had brought me to the United States less than two months before. I would not take the liberty, after so short a period of time, of giving an opinion on America and the Americans. I would only say how deeply I have been touched by the warm reception which was tendered everywhere to me and my daughters. Our hosts wanted to make us feel that we were not with strangers; and, on the other hand, many of them assured me that they felt in entirely friendly surroundings when on the soil of France. I got back to France with a feeling of gratitude for the precious gift of the American women, and with a feeling of affection for their great country tied with ours by a mutual sympathy which gives confidence in a peaceful future for humanity.

[15]Quite recently there has been started near Anverst an important radium industry as a result of the discovery of uranium ore in the Belgian Congo.

# THE DISCOVERY OF RADIUM

(Vassar lecture)

I could tell you many things about radium and radioactivity and it would take a long time. But as we can not do that, I shall only give you a short account of my early work about radium. Radium is no more a baby, it is more than twenty years old, but the conditions of the discovery were somewhat peculiar, and so it is always of interest to remember them and to explain them.

We must go back to the year 1897. Professor Curie and I worked at that time in the laboratory of the school of Physics and Chemistry where Professor Curie held his lectures. I was engaged in some work on uranium rays which had been discovered two years before by Professor Becquerel. I shall tell you how these uranium rays may be detected. If you take a photographic plate and wrap it in black paper and then on this plate, protected from ordinary light, put some uranium salt and leave it a day, and the next day the plate is developed, you notice on the plate a black spot at the place where the uranium salt was. This spot has been made by special rays which are given out by the uranium and are able to make an impression on the plate in the same way as ordinary light. You can also test those rays in another way, by placing them on an electroscope. You know what an electroscope is. If you charge it, you can keep it charged several hours and more, unless uranium salts are placed near to it. But if this is the case the electroscope loses its charge and the gold or aluminum leaf falls gradually in a progressive way. The speed with which the leaf moves may be used as a measure of the intensity of the rays; the greater the speed, the greater the intensity.

I spent some time in studying the way of making good measurements of the uranium rays, and then I wanted to know if there were other elements, giving out rays of the same kind. So I took up a work about all known elements, and their compounds and found that uranium compounds are active and also all thorium compounds, but other elements were not found active, nor were their compounds. As for the uranium and thorium compounds, I found that they were active in proportion to their uranium or thorium content. The more uranium or thorium, the greater the activity, the activity being an atomic property of the elements, uranium and thorium.

Then I took up measurements of minerals and I found that several of those which contain uranium or thorium or both were active. But then the activity was not what I could expect, it was greater than for uranium or thorium compounds like the oxides which are almost entirely composed of these elements. Then I thought that there should be in the minerals some unknown element having a much greater radioactivity than uranium or thorium. And I wanted to find and to separate that element, and I settled to that work with Professor Curie. We thought it would be done in several weeks or months, but it was not so. It took many years of hard work to finish that task. There was not one new element, there were several of them. But the most important is radium which could be separated in a pure state.

All the tests for the separation were done by the method of electrical measurements with some kind of electroscope. We just had to make chemical separations and to examine all products obtained with respect to their activity. The product which retained the radioactivity was considered as that one which had kept the new element; and, as the radioactivity was more strong in some products, we knew that we had succeeded in concentrating the new element. The radioactivity was used in the same way as a spectroscopical test.

The difficulty was that there is not much radium in a mineral; this we did not know at the beginning. But we now know that there is not even one part of radium in a million parts of good ore. And too, to get a small quantity of pure radium salt, one is obliged to work up a huge quantity of ore. And that was very hard in a laboratory.

We had not even a good laboratory at that time. We worked in a hangar where there were no improvements, no good chemical arrangements. We had no help, no money. And because of that the work could not go on as it would have done under better conditions. I did myself the numerous crystalizations which were wanted to get the radium salt separated from the barium salt with which it is obtained out of the ore. And in 1902 I finally succeeded in getting pure radium chloride and determining the atomic weight of the new element radium, which is 226 while that of barium is only 137.

Later I could also separate the metal radium, but that was a very difficult work; and, as it is not necessary for the use of radium to have it in this state, it is not generally prepared that way.

Now, the special interest of radium is in the intensity of its rays which is several million times greater than the uranium rays. And the effects of the rays make the radium so important. If we take a practical point of view, then the most important property of the rays is the production of physiological effects on the cells of the human organism. These effects may be used for the cure of several diseases. Good results have been obtained in many cases. What is considered particularly important is the treatment of cancer. The medical utilization of radium makes it necessary to get that element in sufficient quantities. And so a factory of radium was started to begin with in France, and later in America where a big quantity of ore named carnotite is available. America does produce many grams of radium every year but the price is still very high because the quantity of radium contained in the ore is so small. The radium is more than a hundred thousand times dearer than gold.

But we must not forget that when radium was discovered no one knew that it would prove useful in hospitals. The work was one of pure science. And this is a proof that scientific work must not be considered from the point of view of the direct usefulness of it. It must be done for itself, for the beauty of science, and then there is always the chance that a scientific discovery may become like the radium a benefit for humanity.

But science is not rich, it does not dispose of important means, it does not generally meet recognition before the material usefulness of it has been proved. The factories produce many grams of radium every year, but the laboratories have very small quantities. It is the same for my laboratory and I am very grateful to the American women who wish me to have more of radium and give me the opportunity of doing more work with it.

The scientific history of radium is beautiful. The properties of the rays have been studied very closely. We know that particles are expelled from radium with a very great velocity near to that of the light. We know that the atoms of radium are destroyed by expulsion of these particles, some of which are atoms of helium. And in that way it has been proved that the radioactive elements are constantly disintegrating and that they produce at the end ordinary elements, principally helium and lead. That is, as you see, a theory of transformation of atoms which are not stable, as was believed before, but may undergo spontaneous changes.

Radium is not alone in having these properties. Many having other radioelements are known already, the polonium, the mesothorium, the radiothorium, the actinium. We know also radioactive gases, named emanations. There is a great variety of substances and effects in radioactivity. There is always a vast field left to experimentation and I hope that we may have some beautiful progress in the following years. It is my earnest desire that some of you should carry on this scientific work and keep for your ambition the determination to make a permanent contribution to science.

M. Curie.

*With my friendship for the Students of Vassar College—*

*M. Curie*

# NOBEL PRIZE LECTURE

Nobel Lecture, December 11, 1911*

RADIUM AND THE NEW CONCEPTS IN CHEMISTRY

Some 15 years ago the radiation of uranium was discovered by Henri Becquerel[1], and two years later the study of this phenomenon was extended to other substances, first by me, and then by Pierre Curie and myself[2]. This study rapidly led us to the discovery of new elements, the radiation of which, while being analogous with that of uranium, was far more intense. All the elements emitting such radiation I have termed radioactive, and the new property of matter revealed in this emission has thus received the name radioactivity. Thanks to this discovery of new, very powerful radioactive substances, particularly radium, the study of radioactivity progressed with marvellous rapidity: Discoveries followed each other in rapid succession, and it was obvious that a new science was in course of development. The Swedish Academy of Sciences was kind enough to celebrate the birth of this science by awarding the Nobel Prize for Physics to the first workers in the field, Henri Becquerel, Pierre Curie and Marie Curie (1903).

From that time onward numerous scientists devoted themselves to the study of radioactivity. Allow me to recall to you one of them who, by the certainty of his judgement, and the boldness of his hypotheses and through the many investigations carried out by him and his pupils, has succeeded not only in increasing our knowledge but also in classifying it with great clarity; he has provided a backbone for the new science, in the form of a very precise theory admirably suited to the study of the phenomena. I am happy to recall that Rutherford came to Stockholm in 1908 to receive the Nobel Prize as a well-deserved reward for his work.

Far from halting, the development of the new science has constantly continued to follow an upward course. And now, only 15 years after Becquerel's discovery, we are face to face with a whole world of new phenomena belonging to a field which, despite its close connexion with the fields of physics and chemistry, is particularly well-defined. In this field the importance of radium from, the viewpoint of general theories has been decisive. The history of the discovery and the isolation of this substance has furnished proof of my hypothesis that radioactivity is an atomic property of matter and can provide a means of seeking new elements. This hypothesis has led to present-day theories of radioactivity, according to which we can predict with certainty the existence of about 30 new elements which we cannot generally either isolate or characterize by chemical methods. We also assume that these elements undergo atomic transformations, and the most direct proof in favour of this theory is provided by the experimental fact of the formation of the chemically defined element helium starting from the chemically-defined element radium.

Viewing the subject from this angle, it can be said that the task of isolating radium is the corner-stone of the edifice of the science of radioactivity. Moreover, radium remains the most useful and powerful tool in radioactivity laboratories. I believe that it is because of these considerations that the Swedish Academy of Sciences has done me the very great honour of awarding me this year's Nobel Prize for Chemistry.

It is therefore my task to present to you radium in particular as a new chemical element, and to leave aside the description of the many radioactive phenomena which have already been described in the Nobel Lectures of H. Becquerel, P. Curie and E. Rutherford.

Before broaching the subject of this lecture, I should like to recall that the discoveries of radium and of polonium were made by Pierre Curie in collaboration with me. We are also indebted to Pierre Curie for basic research in the field of radioactivity, which has been carried out either alone, in collaboration with his pupils.

The chemical work aimed at isolating radium in the state of the pure salt, and at characterizing it as a new element, was carried out specially by me, but it is intimately connected with our common work.

I thus feel that I interpret correctly the intention of the Academy of Sciences in assuming that the award of this high distinction to me is motivated by this common work and thus pays homage to the memory of Pierre Curie.

I will remind you at the outset that one of the most important properties of the radioactive elements is that of ionizing the air in their vicinity (Becquerel). When a uranium compound is placed on a metal plate A situated opposite another plate B and a difference in potential is maintained between the plates A and B, an electric current is set up between these plates; this current can be measured with accuracy under suitable conditions and will serve as a measure of the activity of the substance. The conductivity imparted to the air can be ascribed to ionization produced by the rays emitted by the uranium compounds.

In 1897, using this method of measurement, I undertook a study of the radiation of uranium compounds, and soon extended this study to other substances, with the aim of finding out whether radiation of this type occurs in other elements. I found in this way that of the other elements known, only the compounds of thorium behave like the compounds of uranium.

I was struck by the fact that the activity of uranium and thorium compounds appears to be an atomic property of the element uranium and of the element thorium. Chemical compounds and mixtures containing uranium and thorium are active in direct proportion to the amount of these metals contained in them. The activity is not destroyed by either physical changes of state or chemical transformations.

I measured the activity of a number of minerals; all of them that appear to be radioactive always contain uranium or thorium. But an unexpected fact was noted: certain minerals (pitchblende, chalcolite, autunite) had a greater activity than might be expected on the basis of their uranium or thorium content. Thus, certain pitchblendes containing 75% of uranium oxide are about four times as radioactive as this oxide. Chalcolite (crystallized phosphate of copper and uranium) is about twice as radioactive as uranium. This conflicted with views which held that no mineral should be more radioactive than metallic uranium. To explain this point I prepared synthetic chalcolite from pure products, and obtained crystals, whose activity was completely consistent with their uranium content; this activity is about half that of uranium.

I then thought that the greater activity of the natural minerals might be determined by the presence of a small quantity of a highly-radioactive material, different from uranium, thorium and the elements known at present. It also occurred to me that if this was the case I might be able to extract this substance from the mineral by the ordinary methods of chemical analysis. Pierre Curie and I at once carried out this research, hoping that the proportion of the new element might reach several per cent. In reality the proportion of the hypothetical element was far lower and it took several years to show unequivocally that pitchblende contains at least one highly-radioactive material which is a new element in the sense that chemistry attaches to the term.

We were thus led to create a new method of searching for new elements, a method based on radioactivity considered as an atomic property of matter. Each chemical separation is followed by a measurement of the activity of the products obtained, and in this way it is possible to determine how the active substance behaves from the chemical viewpoint. This method has come into general application, and is similar in some ways to spectral analysis. Because of the wide variety of radiation emitted, the method could be perfected and extended, so that it makes it possible, not only to discover radioactive materials, but also to distinguish them from each other with certainty.

It was also found in using the method being considered, that it was in fact possible to concentrate the activity by chemical methods. We found that pitchblende contains at least two radioactive materials, one of which, accompanying bismuth, has been given the name polonium, while the other, paired with barium, has been called radium.

Other radioactive elements have been discovered since: actinium (Debierne), radiothorium and mesothorium (Hahn), ionium (Boltwood), etc.

We were convinced that the materials which we had discovered were new chemical elements. This conviction was based solely on the atomic nature of radioactivity. But at first, from the chemical viewpoint, it was as if our substances had been, the one pure bismuth, and the other pure barium. It was vital to show that the radioactive property was connected with traces of elements that were neither bismuth nor barium. To do that the hypothetical elements had to be isolated. In the case of radium isolation was completely successful but required several years of unremitting effort. Radium in the pure salt form is a substance the manufacture of which has now been industrialized; for no other new radioactive substance have such positive results been obtained.

The radiferous minerals are being subjected to very keen study because the presence of radium lends them considerable value. They are identifiable either by the electrometric method, or very simply by the impression they produce on a photographic plate. The best radium mineral is the pitchblende from St. Joachimsthal (Austria) which has for a long time been processed to yield uranium salts. After extraction of the latter, the mineral leaves a residue which contains radium and polonium. We have normally used this residue as our raw material.

The first treatment consists in extracting the radiferous barium and the bismuth containing the polonium. This treatment, which was first performed in the laboratory on several kilograms of raw material (as many as 20 kg) had then to be undertaken in a factory owing to the need to process thousands of kilograms. Actually, we gradually learned from experience that the radium is contained in the raw material in the proportion of a few decigrams per ton. About 10 to 20 kg crude barium sulphate containing radium are extracted from one ton of residue. The activity of these sulphates is even then 30 to 60 times greater than that of uranium. These sulphates are purified and converted to chlorides. In the mixture of barium and radium chlorides the radium is present only in the proportion of about 3 parts per

100,000. In the radium industry in France a much lower grade mineral is most often used and the proportion indicated is far lower still. To separate the radium from the barium I have used a method of fractional crystallization of the chloride (the bromide can also be used). The radium salt, less soluble than the barium salt, becomes concentrated in the crystals. Fractionation is a lengthy, methodical operation which gradually eliminates the barium. To obtain a very pure salt I have had to perform several thousands of crystallizations. The progress of the fractionation is monitored by activity measurements.

A first proof that the element radium existed was furnished by spectral analysis. The spectrum of a chloride enriched by crystallization exhibited a new line which Demarcay attributed to the new element. As the activity became more concentrated, the new line increased in intensity and other lines appeared while the barium spectrum became at the same time less pronounced. When the purity is very high the barium spectrum is scarcely visible.

I have repeatedly determined the average atomic weight of the metal in the salt subjected to spectral analysis. The method used was the one consisting in determining the chlorine content in the form of silver chloride in a known amount of the anhydrous chloride. I have found that this method gives very good results even with quite small amounts of substance (0.1 to 0.5 g), provided a very fast balance is used to avoid the absorption of water by the alkaline-earth salt during the weighings. The atomic weight increases with the enrichment of the radium as indicated by the spectrum. The successive atomic weights obtained were: 138; 146; 174; 225; 226.45. This last value was determined in 1907 with 0.4 g of very pure radium salt. The results of a number of determinations are, 226.62; 226.31; 226.42. These have been confirmed by more recent experiments.

The preparation of pure radium salts and the determination of the atomic weight of radium have proved positively that radium is a new element and have enabled a definite position to be assigned to it. Radium is the higher homologue of barium in the family of alkaline-earth metals; it has been entered in Mendeleev's table in the corresponding column, on the row containing uranium and thorium. The radium spectrum is very precisely known. These very clear-cut results for radium have convinced chemists and justified the establishment of the new science of radioactive substances.

In chemical terms radium differs little from barium; the salts of these two elements are isomorphic, while those of radium are usually less soluble than the barium salts. It is very interesting to note that strong radioactivity of radium involves no chemical anomalies and that the chemical properties are actually those which correspond to the position in the Periodic System indicated by its atomic weight. The radioactivity of radium in solid salts is ca. 5 million times greater than that of an equal weight of uranium. Owing to this activity its salts are spontaneously luminous. I also wish to recall that radium gives rise to a continuous liberation of energy which can be measured as heat, being about 118 calories per gram of radium per hour.

Radium has been isolated in the metallic state (M. Curie and A. Debierne, 1910). The method used consisted in distilling under very pure hydrogen the amalgam of radium formed by the electrolysis of a chloride solution using a mercury cathode. One decigram only of salt was treated and consequently considerable difficulties were involved. The metal obtained melts at about 700°C, above which temperature it starts to volatilize. Is it very unstable in the air and decomposes water vigorously.

The radioactive properties of the metal are exactly the ones that can be forecast on the assumption that the radioactivity of the salts is an atomic property of the radium which is unaffected by the state of combination. It was of real importance to corroborate this point as misgivings had been voiced by those to whom the atomic hypothesis of radioactivity was still not evident.

Although radium has so far only been obtained in very small amounts, it is nevertheless true to say, in conclusion, that it is a perfectly defined and already well-studied chemical element.

Unfortunately, the same cannot be stated for polonium, for which nevertheless considerable effort has already been spent. The stumbling block here is the fact that the proportion of polonium in the mineral is about 5,000 times smaller than that of radium.

Before theoretical evidence was available from which to forecast this proportion, I had conducted several extremely laborious operations to concentrate polonium and in this way had secured products with very high activity without being able to arrive at definite results as in the case of radium. The difficulty is heightened by the fact that polonium disintegrates spontaneously, disappearing by half in a period of 140 days. We now know that radium has not an infinite life either, but the rate of disappearance is far less (it disappears by half in 2,000 years). With our facilities we can scarcely hope to determine the atomic weight of polonium because theory foresees that a rich mineral can contain only a few hundredths of a milligram per ton, but we can hope to observe its spectrum. The operation of concentrating polonium, as I shall point out later, is, moreover, a problem of great theoretical interest.

Recently, in collaboration with Debierne, I undertook to treat several tons of residues from uranium mineral with a view to preparing polonium. Initially conducted in the factory, then in the laboratory, this treatment finally yielded a few milligrams of substance about 50 times more active than an equal weight of pure radium. In the spectrum of the substance some new lines could be observed which appear attributable to polonium and of which the most important has the wavelength 4170.5 Å. According to the atomic hypothesis of radioactivity, the polonium spectrum should disappear at the same time as the activity and this fact can be confirmed experimentally,

I have so far considered radium and polonium only as chemical substances. I have shown how the fundamental hypothesis which states that radioactivity is an atomic property of the substance has led to the discovery of new chemical elements. I shall now describe how the scope of this hypothesis has been greatly enlarged by the considerations and experimental facts which resulted in establishing the theory of atomic radioactive transformations.

The starting-point of this theory must be sought in the considerations of the source of the energy involved in the phenomena of radioactivity. This energy becomes manifest as an emission of rays which produce thermal, electrical and light phenomena. As the emission occurs spontaneously without any known cause of excitation, various hypotheses have been advanced to account for the liberation of energy. One of the hypotheses put forward at the beginning of our research by Pierre Curie and myself consisted in assuming that the radiation is an emission of matter accompanied by a loss in weight of the active substances and that the energy is taken from the substance itself whose evolution is not yet completed and which undergoes an atomic transformation. This hypothesis, which at first could only be enunciated together with other equally valid theories, has attained dominant importance and finally asserted itself in our minds owing to a body of experimental evidence which substantiated it. This evidence is essentially the following: A series of radioactive phenomena exists in which radioactivity appears to be tied up to matter in an imponderable quantity, the radiation moreover not being permanent but disappearing more or less rapidly with time. Such are polonium, radioactive emanations and deposits of induced radioactivity.

It has been established moreover in certain cases that the radioactivity observed increases with time. This is what happens in the case of freshly prepared radium, of the emanation freshly introduced into the measuring apparatus, of thorium deprived of thorium X, etc.

A careful study of these phenomena has shown that a very satisfactory general explanation can be given by assuming that each time a decrease of radioactivity is observed there is a destruction of radioactive matter, and that each time an increase of activity is observed, there is a production of radioactive matter. The radiations which disappear and appear are, besides, of very varied nature and it is admitted that every kind of rays determined can serve to characterize a substance which is its source, and appears and disappears with it.

As radioactivity is in addition a property which is essentially atomic, the production or the destruction of a distinct type of radioactivity corresponds to a production or a destruction of atoms of a radioactive substance.

Finally, if it is supposed that radioactive energy is a phenomenon which is borrowed from atomic transformation, it can be deduced from this that every radioactive substance undergoes such a transformation, even though it appears to us to be invariable. Transformation in this case is only very slow and this is what takes place in the case of radium or uranium.

The theory I have just summarized is the work of Rutherford and Soddy, which they have called theory of atomic disintegration. By applying this theory it can be concluded that a primary radioactive substance such as radium undergoes a series of atomic transmutations by virtue of which the atom of radium gives birth to a train of atoms of smaller and smaller weights, since a stable state cannot be attained as long as the atom formed is radioactive. Stability can only be attained by inactive matter.

From this point of view one of the most brilliant triumphs of the theory is the prediction that the gas helium, always present in radioactive minerals, can represent one of the end-products of the evolution of radium, and that it is in the form of alpha rays that the helium atoms which are formed when radium atoms distintegrate are discharged. Now, the production of helium by radium has been proved by the experiments of Ramsay and Soddy, and it cannot now be contested that the perfectly defined chemical element, radium, gives rise to the formation of another equally defined element – helium. Moreover, the investigations done by Rutherford and his students have proved that the alpha particles emitted by radium with an electric charge are also to be found in the form of helium gas in the space where they have been recovered.

I must remark here that the bold interpretation of the relationship existing between radium and helium rests entirely upon the certitude that radium has the same claim to be a chemical element as have all the other known elements, and that there can be no question of regarding it to be a molecular combination of helium with another element. This shows how fundamental in these circumstances has been the work carried out to prove the chemical individuality of radium, and it can also be seen in what way the hypothesis of the atomic nature of radioactivity and the theory of radioactive transformations have led to the experimental discovery of a first clearly-established example of atomic transmutation. This is a fact the significance of which cannot escape anyone, and one which incontestably marks an epoch from the point of view of chemists.

Considerable work, guided by the theory of radioactive transformations, has led to approximately 30 new radioactive elements being envisaged, classified in 4 series according to the primary substance: these series are uranium, radium, thorium and actinium. The uranium and radium series can, in fact, be combined, for it seems to be proved that radium is a derivative of uranium. In the radium series the last known radioactive body is polonium, the production of which by radium is now a proven fact. It is likely that the actinium series is related to that of radium.

We have seen that helium gas is one of the products of radium distintegration. The helium atoms are detached from those of radium and its derivatives during the course of the transformation. It is supposed that after the departure of four atoms of helium, the radium atom yields one atom of polonium; the departure of a fifth helium atom determines the formation of an inactive body with an atomic weight believed to be equal to 206 (20 units below that of radium). According to Rutherford this final element is nothing more than lead, and this supposition is now being subjected to experimental verification in my laboratory. The production of helium from polonium has been directly proved by Debierne.

The relatively large amount of polonium prepared by Curie and Debierne has allowed an important study to be undertaken. This consists in counting a large number of alpha particles emitted by polonium and in collecting and measuring the corresponding volume of helium. Since each particle is a helium atom, the number of helium atoms is thus found which occupy a given volume and have a given weight. It can therefore allow us to deduce, in a general way, the number of molecules in a grammolecule. This number, known as Avogadro's constant, is of great importance. Experiments conducted on polonium have supplied a first value for this number, which is in good agreement with the values obtained by other methods. The enumeration of alpha particles is done by an electrometric method due to Rutherford; this method has been brought to perfection by means of a photographic recording apparatus.

Recent investigations have shown that potassium and rubidium emit a very feeble radiation, similar to the beta radiation of uranium and radium. We do not yet know whether we should regard these substances as true radioactive bodies, i.e. bodies in the process of transformation.

To conclude I should like to emphasize the nature of the new chemistry of radioactive bodies. Tons of material have to be treated in order to extract radium from the ore. The quantities of radium available in a laboratory are of the order of one milligram, or of a gram at the very most, this substance being worth 400,000 francs per gram. Very often material has been handled in which the presence of radium could not be detected by the balance, nor even by the spectroscope. And yet we have methods of measuring so perfect and so sensitive that we are able to know very exactly the small quantities of radium we are using. Radioactive analysis by electrometric methods allows us to calculate to within 1% a thousandth of a milligram of radium, and to detect the presence of 10-10 grams of radium diluted in a few grams of material. This method is the only one which could have led to the discovery of radium in view of the dilution of this substance in the ore. The sensitivity of the methods is still more striking in the case of radium emanation, which can be detected when the quantity present amounts, for example, to only 10-10 mm[3]. As the specific activity of a substance is, in the case of analogous radiations, approximately in inverse proportion to the average life, the result is that if the average life is very brief, the radioactive reaction can attain an unprecedented sensitivity. We are also accustomed to deal currently in the laboratory with substances the presence of which is only shown to us by their radioactive properties but which nevertheless we can determine, dissolve, reprecipitate from their solutions and deposit electrolytically. This means that we have here an entirely separate kind of chemistry for which the current tool we use is the electrometer, not the balance, and which we might well call the chemistry of the imponderable.

*The lecture was held in the lecture hall at the Royal Academy of Sciences. The Royal Academy of Sciences were at the time located in the Westman Palace, a building behind Adolf Fredrik church in Stockholm, Sweden.

1. H. Becquerel, Compt. Rend., (1896).

2. P. Curie and M. Curie, Compt. Rend., (1898); (1899).

3. M. Curie, Rev. Gen. Sci., (1899); Rev. Sci., (1900).

www.ingramcontent.com/pod-product-compliance
Lightning Source LLC
LaVergne TN
LVHW022222200926
848481LV00006B/2083